HIGH-DENSITY INTEGRATED ELECTROCORTICAL NEURAL INTERFACES

HIGH-DENSITY INTEGRATED ELECTROCORTICAL NEURAL INTERFACES

Low-Noise Low-Power System-on-Chip Design Methodology

SOHMYUNG HA
Division of Engineering
New York University Abu Dhabi
Abu Dhabi, United Arab Emirates

CHUL KIM
Department of Bio and Brain Engineering
Korea Advanced Institute of Science and Technology (KAIST)
Daejeon, South Korea

PATRICK P. MERCIER
Department of Electrical and Computer Engineering
University of California, San Diego
La Jolla, CA, United States

GERT CAUWENBERGHS
Department of Bioengineering
University of California, San Diego
La Jolla, CA, United States

ACADEMIC PRESS
An imprint of Elsevier

Academic Press is an imprint of Elsevier
125 London Wall, London EC2Y 5AS, United Kingdom
525 B Street, Suite 1650, San Diego, CA 92101, United States
50 Hampshire Street, 5th Floor, Cambridge, MA 02139, United States
The Boulevard, Langford Lane, Kidlington, Oxford OX5 1GB, United Kingdom

Notices

Knowledge and best practice in this field are constantly changing. As new research and experience broaden our
understanding, changes in research methods, professional practices, or medical treatment may become necessary.

Practitioners and researchers must always rely on their own experience and knowledge in evaluating and using any
information, methods, compounds, or experiments described herein. In using such information or methods they
should be mindful of their own safety and the safety of others, including parties for whom they have a professional
responsibility.

To the fullest extent of the law, neither the Publisher nor the authors, contributors, or editors, assume any liability
for any injury and/or damage to persons or property as a matter of products liability, negligence or otherwise, or
from any use or operation of any methods, products, instructions, or ideas contained in the material herein.

Library of Congress Cataloging-in-Publication Data
A catalog record for this book is available from the Library of Congress

British Library Cataloguing-in-Publication Data
A catalogue record for this book is available from the British Library

ISBN: 978-0-12-815115-0

For information on all Academic Press publications
visit our website at https://www.elsevier.com/books-and-journals

Publisher: Mara Conner
Acquisition Editor: Chris Katsaropoulos
Editorial Project Manager: Gabriela D. Capille
Production Project Manager: R. Vijay Bharath
Designer: Miles Hitchen

Typeset by VTeX

Working together
to grow libraries in
developing countries

www.elsevier.com • www.bookaid.org

Contents

About the authors

Sohmyung Ha is an Assistant Professor of Electrical Engineering at New York University Abu Dhabi, Abu Dhabi, UAE, and a Global Network Assistant Professor of Electrical and Computer Engineering at New York University, Brooklyn, NY, USA. He received the BS (*summa cum laude*) and MS degrees in electrical engineering from the Korea Advanced Institute of Science and Technology (KAIST), Daejeon, Korea, in 2004 and 2006, respectively. From 2006 to 2010, he worked at Samsung Electronics as analog and mixed-signal circuit designer for commercial multimedia devices. After this extended career in the industry, he returned to academia as a Fulbright Scholar, and obtained the MS and PhD degrees in Bioengineering with the best PhD thesis award for biomedical engineering from the Department of Bioengineering, University of California San Diego, La Jolla, CA, USA, in 2015 and 2016, respectively. Since 2016, he has been with New York University Abu Dhabi and New York University.

Dr. Ha has served as an Associate Editor of Smart Health (Elsevier) since 2016 and IEEE Transactions on Biomedical Circuits and Systems since 2019, and is currently a member of the Analog Signal Processing Technical Committee of IEEE Circuits and Systems Society. His research aims at advancing the engineering and applications of silicon integrated technology interfacing with biology in a variety of forms, ranging from implantable biomedical devices to unobtrusive wearable sensors.

Chul Kim is an Assistant Professor of Bio and Brain Engineering at the Korea Advanced Institute of Science and Technology (KAIST), Daejeon, South Korea. He received the BS degree in electrical engineering from the Kyungpook National University, Daegu, Korea, in 2007, the MS degree in electrical engineering from KAIST, in 2009, and the PhD degree in bioengineering from the UC San Diego, La Jolla, CA, USA, in 2017.

From 2009 to 2012, he was with SK HYNIX, Icheon, Korea, where he designed power management circuitry for dynamic random access memory. His current research interests include ultra low-power integrated circuits and systems for fully wireless brain–machine interfaces and body-area networks.

Dr. Kim received Gold prize in the 16th Human-Tech thesis prize contest from Samsung Electronics, Suwon, South Korea, in 2010 and the 2017–2018 Shunichi Usami PhD Thesis Design Award in the Bioengineering Department, UC San Diego, La Jolla, CA, USA. He was the recipient of the 2017–2018 IEEE Solid-State Circuits Society Predoctoral Achievement Award.

Patrick P. Mercier received the BSc degree in electrical and computer engineering from the University of Alberta, Edmonton, AB, Canada, in 2006, and the SM and PhD degrees in electrical engineering and computer science from the Massachusetts Institute of Technology (MIT), Cambridge, MA, USA, in 2008 and 2012, respectively. He is currently an Associate Professor in Electrical and Computer Engineering at the University of California San Diego (UCSD), where he is also the co-director of the Center for Wearable Sensors. His research interests include the design of energy-efficient microsystems, focusing on the design of RF circuits, power converters, and sensor interfaces for miniaturized systems and biomedical applications.

Prof. Mercier received a Natural Sciences and Engineering Council of Canada (NSERC) Julie Payette fellowship in 2006, NSERC Postgraduate Scholarships in 2007 and 2009, an Intel PhD Fellowship in 2009, the 2009 IEEE International Solid-State Circuits Conference (ISSCC) Jack Kilby Award for Outstanding Student Paper at ISSCC 2010, a Graduate Teaching Award in Electrical and Computer Engineering at UCSD in 2013, the Hellman Fellowship Award in 2014, the Beckman Young Investigator Award in 2015, the DARPA Young Faculty Award in 2015, the UC San Diego Academic Senate Distinguished Teaching Award in 2016, the Biocom Catalyst Award in 2017, and the NSF CAREER Award in 2018. He has served as an Associate Editor of the IEEE Transactions on Very Large Scale Integration during 2015–2017. Since 2013, he has served as an Associated Editor of the IEEE Transactions on Biomedical Circuits and Systems, and is currently a member of the ISSCC International Technical Program Committee (Technology Directions Sub-Committee), the CICC Technical Program Committee, the VLSI Symposium Technical Program Committee, and an Associate Editor of the IEEE Solid-State Circuits Letters. Prof. Mercier was the co-editor of Ultra-Low-Power Short Range Radios (Springer, 2015) and Power Management Integrated Circuits (CRC Press, 2016).

Gert Cauwenberghs received the MEng degree in applied physics from the University of Brussels, Brussels, Belgium, in 1988, and the MS and PhD degrees in electrical engineering from the California Institute of Technology, Pasadena, in 1989 and 1994, respectively.

Currently, he is a Professor of Bioengineering at the University of California, San Diego, where he co-directs the Institute for Neural Computation, participates as a member of the Institute of Engineering in Medicine, and serves on the Computational Neuroscience Executive Committee of the Department of Neurosciences graduate program. Previously, he was a Professor of Electrical and Computer Engineering at Johns Hopkins University, Baltimore, MD, and a Visiting Professor of Brain and Cognitive Science at the Massachusetts Institute of Technology, Cambridge. He has co-founded and chairs the Scientific Advisory Board of Cognionics Inc. His research focuses on micropower biomedical instrumentation, neuron–silicon and brain–machine interfaces, neuromorphic engineering, and adaptive intelligent systems.

Dr. Cauwenberghs received the National Science Foundation Career Award in 1997, the ONR Young Investigator Award in 1999, and the Presidential Early Career Award for Scientists and Engineers in 2000. He was Francqui Fellow of the Belgian American Educational Foundation. He was a Distinguished Lecturer of the IEEE Circuits and Systems Society during 2002–2003. He served IEEE in a variety of roles including as General Chair of the IEEE Biomedical Circuits and Systems Conference (BioCAS 2011, San Diego), as Program Chair of the IEEE Engineering in Medicine and Biology Conference (EMBC 2012, San Diego), and as Editor-in-Chief of the IEEE Transactions on Biomedical Circuits and Systems.

Preface

This book aims to provide a basic understanding, design strategies, and implementation applications for electrocortical neural interfaces with a focus on integrated circuit design technologies. A wide variety of topics associated with the design and application of electrocortical neural implants are covered. It discusses basic principles and practical design strategies of electrocorticography, electrode interfaces, signal acquisition, power delivery, data communication, and stimulation. In addition, an overview and critical review of the state-of-the-art are included. These methodologies present a path towards the development of minimally invasive brain–computer interfaces capable of resolving microscale neural activity with wide-ranging coverage across the cortical surface.

Sohmyung Ha, Chul Kim, Patrick P. Mercier, and Gert Cauwenberghs
Abu Dhabi, UAE, Daejeon, South Korea, and La Jolla, CA, USA
March 15, 2019

Acknowledgment

We thank Gabriela Capille, Chris Katsaropoulos, Vijay Bharath, Sheela Bernardine Josy and their teams including the Design team at Elsevier for professional guidance and kind support throughout the publishing process. We are also grateful to all the anonymous referees for carefully reviewing the manuscript during this process. We also thank Dr. Jacobo Reyes-Velasco at New York University Abu Dhabi for the chip photos for the book cover.

Sohmyung Ha, Chul Kim, Patrick P. Mercier, and Gert Cauwenberghs
March 15, 2019

CHAPTER 1

Introduction to ECoG interfaces

Contents

1.1. Introduction

The Brain Research through Advancing Innovative Neurotechnologies (BRAIN) Initiative envisions expanding our understanding of the human brain. It targets development and application of innovative neural technologies to advance the resolution of neural recording, and stimulation toward dynamic mapping of the brain circuits and processing [1,2]. These advanced neurotechnologies will enable new studies and experiments to augment our current understanding of the brain, thereby enabling tremendous advances in diagnosis and treatment opportunities over a broad range of neurological diseases and disorders.

Studying the dynamics and connectivity of the brain requires a wide range of technologies to address multiple temporal and spatial scales. Fig. 1.1 shows spatial and temporal resolutions and spatial coverage of the various brain monitoring methods that are currently available [3–6].

Noninvasive methods such as magnetic resonance imaging (MRI), functional magnetic resonance imaging (fMRI), magnetoencephalography (MEG), and positron emission tomography (PET) provide whole-brain spatial coverage. Although fMRI achieves high spatial resolution down to 1 mm, its temporal resolution is severely limited (1–10 s) as the system measures neural activity indirectly by quantifying blood oxygenation to

High-Density Integrated Electrocortical Neural Interfaces
https://doi.org/10.1016/B978-0-12-815115-0.00008-8

1

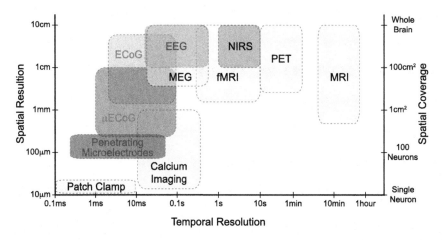

Figure 1.1 Spatial and temporal resolution, as well as spatial coverage, of various neural activity monitoring modalities [4–6]. For each modality shown, the lower boundary of the box specifies the spatial resolution indicated on the left axis, whereas the upper boundary specifies the spatial coverage on the right axis. The width of each box indicates the typical achievable range of temporal resolution. Portable modalities are shown in color. Bridging an important gap between noninvasive and highly invasive techniques, µECoG has emerged as a useful tool for diagnostics and brain-mapping research.

support regions with more elevated metabolism. In contrast, MEG provides higher temporal resolution (0.01–0.1 s) at the expense of poor spatial resolution (1 cm). Whereas fMRI and MEG provide complementary performance in spatiotemporal resolution, PET offers molecular selectivity in functional imaging at the expense of lower spatial (1 cm) and temporal (10–100 s) resolution, and the need for injecting positron emitting radionuclides in the bloodstream. However, neither fMRI, MEG or PET are suitable for wearable or portable applications, as they all require very large, expensive, and high power equipment to support the sensors, as well as extensively shielded environments.

In contrast, electrophysiology methods, which directly measure electrical signals that arise from the activity of neurons, offer superior temporal resolution. They have been extensively used to monitor brain activity due to their ability to capture wide ranges of brain activities from the subcellular level to the whole brain oscillation level as shown in Fig. 1.2. Due to recent advances in electrode and integrated circuit technologies, electrophysiological monitoring methods can be designed to be portable, with fully wearable or implantable configurations for brain–computer interfaces having been demonstrated.

One of the most popular electrophysiological monitoring methods is electroencephalography (EEG), which records electrical activity on the scalp resulting from volume conduction of coherent collective neural activity throughout the brain, as illustrated in Fig. 1.2. EEG recording is safe (noninvasive) and relatively inexpensive, but its spatiotemporal resolution is limited to about 1 cm and 100 Hz, due largely to the dispersive electrical properties of several layers of high-resistive tissue, particularly

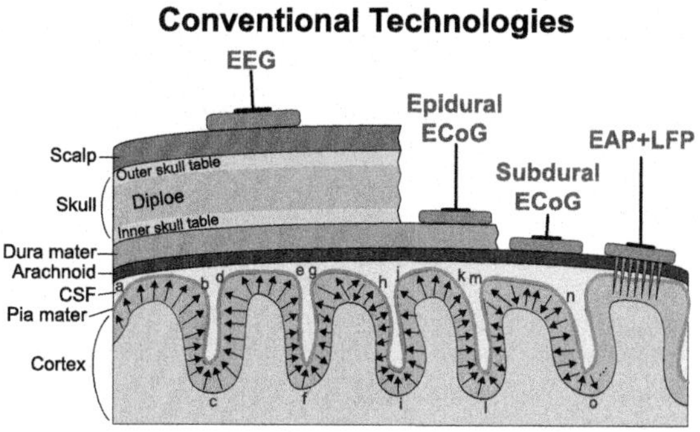

Conventional Technologies

Figure 1.2 Conventional electrophysiology methods including EEG, ECoG and neural spike and LFP recording with penetrating microelectrodes. Both EEG and ECoG can capture correlated collective volume conductions in gyri such as regions of a–b, d–e and j–k. However, they cannot record opposing volume conductions in sulci such as regions of b–c–d and e–f–g and random dipole layers such as regions of g–h and l–m–n–o [17].

skull, between the brain and the scalp. In contrast, recording with intracranial brain-penetrating microelectrodes (labeled as EAP+LFP in Fig. 1.2) can achieve much higher resolution due to the much closer proximity to individual neurons. Thus, it is also widely used for brain research and brain–computer interface (BCI) applications. Using microelectrodes, extracellular action potentials (EAPs) and local field potentials (LFPs) can be recorded from multiple neurons across multiple cortical areas and layers. Even though penetrating microelectrodes can provide rich information from neurons, they can suffer from tissue damage during insertion [7–9], and have substantial limitations in long-term chronic applications due to their susceptibility to signal degradation from electrode displacement and immune response against the electrodes [10]. Because of the more extreme invasiveness and longevity issues, chronic implantation of penetrating microelectrodes in humans is not yet viable.

Between the two extremes of EEG and penetrating microelectrode arrays, a practical alternative technique is electrocoticography (ECoG), or intracranial/intraoperative EEG (iEEG), which records synchronized postsynaptic potentials at locations much closer to the cortical surface, as illustrated in Fig. 1.2. Compared to EEG, ECoG has higher spatial resolution [11–13], higher signal-to-noise ratio, broader bandwidth [14], and much less susceptibility to artifacts from movement, electromyogram (EMG), or electrooculargram (EOG) [15,16]. In addition, ECoG does not penetrate the cortex, does not scar, and can have superior long-term signal stability recording through subdural surface electrodes.

With advances in high channel count and wireless operation, ECoG has recently again emerged as an important tool not only for more effective treatment of epilepsy, but also for investigating other types of brain activity across the cortical surface. ECoG

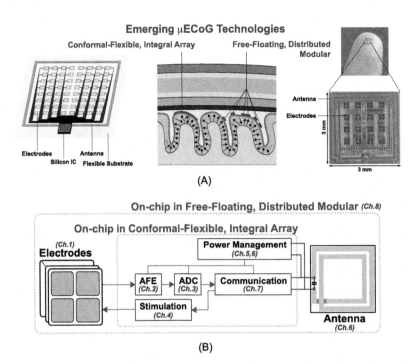

Figure 1.3 (A) Emerging fully implantable μECoG technologies enabled by flexible substrate ECoG microarrays and modular ECoG interface microsystems. Such technologies are capable of capturing local volume conducting activities missed by conventional methods, and are extendable to cover large surface area across cortex. (B) Their functioning block diagram with references to the relevant chapters of this book.

recording provides stable brain activity recording at a mesoscopic spatiotemporal resolution with a large spatial coverage up to whole or a significant area of the brain. Advanced miniaturized electrode arrays have pushed the spatial resolution of ECoG recording to less than 1 mm, offering a unique opportunity to monitor large-scale brain activity much more precisely. Moreover, wireless implantable microsystems based on flexible technology or via modular placement of multi-channel active devices, both illustrated in Fig. 1.3(A), have recently emerged as a new paradigm to record more closely to the cortical surface (in many cases on top of the pia), while enabling coverage along the natural curvature of the cortex without penetration. These micro ECoG, or μECoG, devices enable even higher spatial resolution than conventional ECoG systems, and are beginning to enable next-generation brain mapping, therapeutic stimulation, and BCI systems.

This chapter discusses the general challenges of designing next-generation ECoG interfaces, including recording, miniaturization, stimulation, powering, and data communications. Solutions are presented by surveying the state-of-the-art technologies and example systems. More detailed discussions and implementation can be found in subsequent chapters as indicated in Fig. 1.3(B).

1.2. Electrocorticogram

ECoG recording was pioneered in the 1920s by Hans Berger [18]. He recorded ECoG signals with electrodes placed on the dural surface of human patients. In the 1930s through 1950s, Wilder Penfield and Herbert Jasper at the Montreal Neurological Institute used ECoG to identify epileptogenic zones as a part of the Montreal procedure, which is a surgical protocol to treat patients with severe epilepsy by removing sections of the cortex most responsible for epileptic seizures. In addition, intraoperative electrical stimulation of the brain has been used to explore the functional mapping of the brain including brain areas for speech, motor, and sensory functions. This localization of important brain regions is important to exclude from surgical removal. In spite of recent advances in imaging techniques for functional brain mapping such as fMRI, PET, and MEG, ECoG is still the gold standard for decoding epileptic seizure foci and determining target regions for surgical removal.

The electrocorticogram (also ECoG) is originated from electrical activities of each individual neuron in the brain. Electrical contributions from all transmembrane currents of nearby neurons are superimposed in the extracellular medium, creating combined electric fields in the brain. The electric field can be measured by two or more electrodes placed either on the scalp, on the cortex, or even in the extracellular space inside the brain. The potential recorded on the cortex surface is referred to the ECoG. As the distance between the electrodes and the source gets farther, the contributions from the source becomes less significant and sources nearby contributes more to the recording. Thus, the electric fields from far sources are averaged spatially.

1.3. Volume conduction with differential electrodes

Volume conduction of ionic currents in the brain is the source of ECoG. In the frequency band of interest for ECoG recording (typically less than 500 Hz), the quasi-static electric field equations with conductivities of tissue layers are a good approximate representation [19]. To first order, a volume conducting current monopole I spreads radially through tissue with an outward current density of magnitude $I/4\pi r^2$ at distance r, giving rise to an outward electric field of magnitude $I/4\pi\sigma r^2$ and a corresponding electrical potential $I/4\pi\sigma r$, where σ is the tissue volume conductivity.

1.3.1 Differential recording

For EEG, a current dipole as a closely spaced pair of opposing current monopoles is typically an adequate model representing distant sources of synchronous electrical activity across large assemblies of neurons or synapses [19]. In contrast, for implanted neural recording including ECoG and single-unit neural spike/LFP recording, a set of monopole currents resulting from individual neural units is a more appropriate model at the local spatial scale, especially for high density recording with electrodes spaced at dimensions approaching inter–cellular distances. Since the volume conducting currents from neural action potentials are spatially and temporally distributed, only a few effective current sources at a time are typically active near an electrode, one of which is illustrated in the vicinity of two closely spaced electrodes in Fig. 1.4(A). Furthermore, unlike the ground-referenced recording with single-ended electrodes for EEG, high-density electrode arrays typically require differential recording across electrodes, particularly in μECoG integrated recording since the miniaturized geometry does not allow for a distal ground connection.

In Fig. 1.4(A), the recorded differential voltage V as a function of the distances r_+ and r_- of the two electrodes from a current source I induced by the activity of adjacent neurons can be expressed as:

$$V(r_+, r_-) \cong \frac{I}{2\pi\sigma}\left(\frac{1}{r_+} - \frac{1}{r_-}\right) \tag{1.1}$$

valid for distances r_+, r_- substantially larger than the electrode diameter D. Although the expression is similar to that for an EEG current dipole recorded with a single electrode, it is fundamentally different in that here the difference in monopole activity results from differential sensing with two closely spaced electrodes rather than from dipolar distribution of two closely spaced currents. The factor 2 rather than 4 in the denominator arises from the semi-infinite boundary conditions along the horizontal plane of the electrode substrate, in that volume conduction is restricted to the tissue below the substrate.

Fig. 1.4(B) shows a spatial map of the effect of a current source located in tissue below the electrode pair, with electrode diameter D and pitch $2D$, on the measured differential voltage V. Its recording penetration depth is roughly $2D$, the electrode pitch, vertically, and about $4D$ horizontally. Note again that this is for a single monopolar source; in the presence of dipolar activity with two opposing nearby currents (i.e., charge balancing across a soma and dendrite of a neuron extending below the electrodes) the measured voltage (1.1) becomes double differential, leading to a quadrupolar response profile.

1.3.2 Differential stimulation

The same pair of closely spaced electrodes can be used for differential stimulation by injecting currents into the surrounding tissue. Again, the absence of a distal ground connection in miniature integrated electrode arrays necessitates local charge balancing so that the currents through the two electrodes need to be of equal strength and opposing

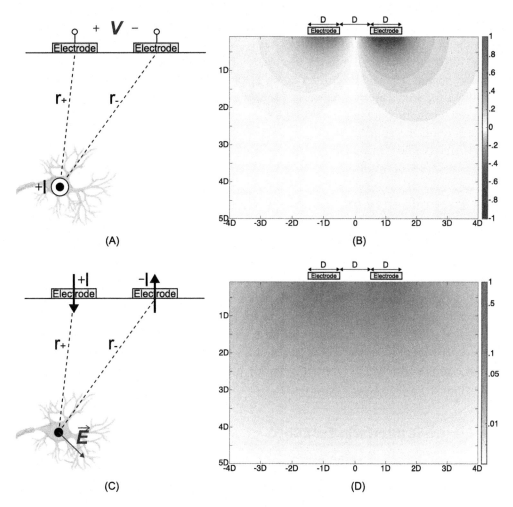

Figure 1.4 (A) Neural recording setting with closely spaced differential electrodes interfacing below with neural tissue of volume conductivity σ. (B) Spatial map of the effect of the location of a current source $+I$ in the tissue on recorded differential voltage V, in units $1/\sigma D$. (C) Neural stimulation setting with differential currents injected into the surrounding tissue through the same two closely spaced electrodes. (D) Spatial map of the resulting electric field magnitude $|\vec{E}|$ in the tissue, in units $1/\sigma D^2$.

polarity, constituting a current dipole sourced within the electrode array. The resulting differential current stimulation can be modeled with the diagram shown in Fig. 1.4(C). The current dipole from the pair of differential stimulation currents flowing through the two electrodes induces an electrical field $\vec{E}$ in the brain tissue expressed by

$$\vec{E}(r_+, r_-) \cong \frac{I}{2\pi\sigma}\left(\frac{\vec{u}_{r_+}}{r_+^2} - \frac{\vec{u}_{r_-}}{r_-^2}\right) \tag{1.2}$$

where again $r_+, r_- \gg D$, and $\vec{\mathbf{u}}_{r_+}$ and $\vec{\mathbf{u}}_{r_-}$ represent unit vectors pointing outwards along the direction of r_+ and r_-, respectively. With the same electrode configuration of Fig. 1.4(B), the magnitude of the electric field $|\vec{\mathbf{E}}|$ is shown in Fig. 1.4(D) indicating a shallow region near the electrodes being electrically stimulated. Note that the electrical field for stimulation is inversely proportional to the square of distance to each electrode, while the potential measured for recording is inversely proportional to linear distance. Thus, the available depth of differential stimulation is shallower than that of differential recording. In general, the penetration depth of stimulation and recording are in the same order of magnitude of the electrode pitch, roughly a few times larger.

1.3.3 Electrode array configurations

While this is a simplified model, it is sufficiently representative to demonstrate the effectiveness of differential electrode configurations for both recording and stimulation without a global reference, as required for fully integrated μECoG in the absence of a distal ground connection. As the analysis and simulations above show, the spatial response of differential recording and stimulation are quite localized near the electrode sites, on a spatial scale that matches the electrode dimensions and spacing. Thus, aside from spatial selection of recording or stimulation along the 2-D surface by translation of selected pairs of adjacent electrodes, depth and spatial resolution of recording or stimulation can be controlled via virtual electrode pitch, by pooling multiple electrodes in complementary pairs of super-electrodes at variable spacing between centers.

1.4. Electrode interfaces for ECoG

Electrodes, which couple ECoG signals from the brain into the analog front-end amplifiers, are the first interface to ECoG systems. Thus, their properties, including materials, geometries, and placement are of crucial importance in building entire acquisition and actuation systems [32].

Given the distance between the scalp and individual neuronal current sources and sinks, EEG recording is unsuitable for detecting small local field potentials as shown in Fig. 1.2. Electrical dipole signals travel a minimum distance of 1 cm between the outer surface of the cerebral cortex to the scalp, including layers of cerebrospinal fluid, meninges, bone and skin, all with varying electrical properties. Through this path the effect of a small-localized dipole source is not only greatly attenuated but also spatially averaged among a myriad of neighbors, resulting in practical and theoretical limits to the spatiotemporal resolution of EEG [33]. As implied in Eq. (1.1), cm-sized electrode arrays with cm spacings in conventional ECoG recordings are better than EEG, but have limitations in resolving current sources of neural activity of size smaller than the electrode pitch. A conventional clinical ECoG array with electrodes at the cm scale is depicted in Fig. 1.5(A).

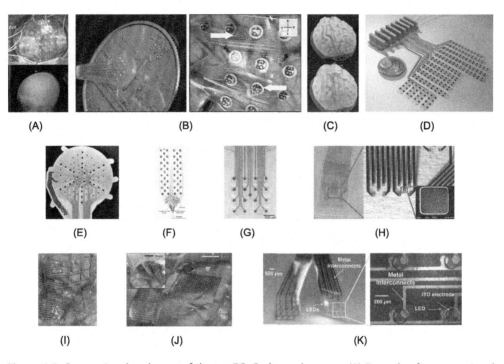

Figure 1.5 Conventional and state-of-the-art ECoG electrode arrays. (A) Example of a conventional electrode array placed on the subdural cortex (top) with post-operative radiograph showing electrode array placement (bottom). The pitch and diameter of electrodes are 1 cm and 2 mm, respectively [20]. (B) μECoG electrode array placed along with a conventional ECoG electrode array [21,22]. (C) Patient-specific electrode array for sulcal and gyral placement [23]. (D) Flexible 252-channel ECoG electrode array on a thin polyimide foil substrate [24]. (E) μECoG electrode array with 124 circular electrodes with three different diameters [25]. (F) Parylene-coated metal tracks and electrodes within a silicone rubber substrate [26]. (G) A transparent μECoG electrode array with platinum electrodes on a parylene-C substrate [27]. (H) An electrode array with poly(3,4-ethylenedioxythiophene) (PDOT) and PEDOT-carbon nanotube (CNT) composite coatings for lower electrode interface impedance [28]. (I) A flexible electrode array on bioresorbable substrates of silk fibroin [29]. (J) Flexible active electrode array with two integrated transistors on each pixel for ECoG signal buffering and column multiplexing for high channel count [30]. (K) Flexible ECoG array with embedded light-emitting diodes for optogenetics-based stimulation [31].

A miniaturized surface electrode array in direct contact with the cerebral cortex can resolve the activity of smaller source populations down to mm or even sub-mm resolution as shown in Fig. 1.5(B)–(K). Furthermore, site specific purposeful electrical stimulation is only possible at μECoG scale. Improvements in the quality and applications of ECoG data have resulted from technological developments at the interface: microfabrication of electrode and substrate materials and interconnect. Simply miniaturizing existing electrode array is not typically sufficient: for example, miniaturized electrodes in <1 mm pitch arrays have very high impedance, which results in poor signal quality, and reduced charge transfer capacity, which typically reduces stimulation efficiency. Such limitations have been addressed by micro-patterning increased surface area, and carbon nanotube [34] or conductive polymer coatings (Fig. 1.5(H)) [28]. Flexible substrates have also reduced the effective distance between source and electrodes through tight, conformal geometries [24,26]. Aside from creating ultra-flexible thin materials, dissolvable substrates leave behind a mesh of thin unobtrusive wires and electrodes with a superior curved conformation and biocompatibility as shown in Fig. 1.5(I) [29,35].

Even with advances in flexible electrode arrays, the number of channels in practical systems are still limited to approximately 100 because of the high density of interconnections between electrode arrays and corresponding acquisition systems. Active electrodes are an emerging approach to maximize number of electrode channels while maintaining small number of wired connections to the electrode array. Advanced fabrication techniques can produce arrays of electrodes with direct integration of transistors on the flexible substrate as shown in Fig. 1.5(J) [30]. This approach can be supplemented with additional in situ devices capable of multiplexing several 100s of recording channels, thereby reducing the required number of wires and interconnections. Another emergent approach combining active recording electrodes and new polymeric materials has led to the development of organic electrochemical transistors in ECoG arrays [36]. However, one limitation of current active arrays is that the same electrode cannot be used for stimulation. Recently, transparent electrode arrays with integrated light path for simultaneous ECoG recording and optogenetic stimulation have been demonstrated as shown in Fig. 1.5(K) [31]. The active development of novel electrode interfaces has not only improved conventional ECoG recording, but also generated new applications and therapeutic opportunities.

1.5. ECoG interfaces: recording and stimulation

1.5.1 Integrated circuit interfaces for data acquisition

Neural data acquisition with a high spatial resolution poses several challenges in the design of application-specific integrated circuits (ASICs) used to perform data acquisition. Higher channel density in ECoG arrays typically results in smaller electrode size, and if the area overhead of the ASIC should be kept small, as is desired in most applications, then the area dedicated to amplify and digitize each channel should also reduce. Unfortunately, area trades off with several important parameters. For example, a more dense array of analog front-end (AFE) amplifiers dissipates more power and generates more heat for each channel, and thus power dedicated to each front-end channel must reduce to meet thermal regulatory limits. Power then trades off with noise, causing signal fidelity issues. The area/volume constraint of front-ends typically also precludes the use of external components such as inductors or capacitors. As a result, AC coupling capacitors employed to reject DC or slowly time-varying electrode offsets are not typically

Table 1.1 Design factors and trade-offs in integrated ECoG interfaces.

Design factors	Trade-offs and inter-relations	Typical range (ECoG)	Examples	Comments
Input referred noise		1–5 μVrms	[45,51,56]	Dominated by electrode interface and front-end amplifier
Power consumption		0.1–10 μW	[46,58,54]	Lowered by biasing, or duty cycling/multiplexing above Nyquist rate
Bandwidth		0.1–500 Hz	[44,69,76]	ECoG signal bands
Dynamic Range (DR)		40–120 dB	[55,67,68]	Variable gain compounds signal-to-noise ratio for extended DR
Variable gain		1–1000 V/V	[42,46,54]	Typically digitally selectable with auto-ranging capability
Power Supply Rejection Ratio (PSRR)		40–80 dB	[42,43,76]	Frequency dependent, limiting switching regulator frequency
Common-Mode Rejection Ratio (CMRR)		60–120 dB	[42,43,45]	Mains interference rejection
Input impedance		100 MΩ–10 TΩ	[48,49,81]	Resistive/capacitive; depending on the contact type and application
DC and low-frequency rejection		0.1–1 Hz	[43,47,49]	AC coupling; with CDS or chopping for 1/f noise and offset reduction
Area / Size		0.1–10 cm^2	[48,69,97]	Dominated by electrodes and battery; minimizing off-chip components

▲: factors to be maximized; ▽: factors to be minimized.

employed, so other techniques are instead necessary. Small electrodes also have higher impedance, requiring even higher AFE input impedance to avoid signal attenuation. In addition, high power supply rejection ratio (PSRR) is required because miniaturized implants typically condition DC power from an external AC source, and further may not be able to accommodate large power decoupling capacitors. Higher channel counts also require higher communication throughput, increasing the power consumption of communication. All these requirements are interrelated and trade off with each other in many ways, as indicated in Table 1.1 [37,32,38].

The noise–current trade-off in instrumentation amplifiers (IAs) is well represented by the noise efficiency factor (NEF) (see Fig. 1.6), which is expressed as

$$NEF = V_{rms,in} \sqrt{\frac{2I_{tot}}{\pi V_t \cdot 4kT \cdot BW}} \tag{1.3}$$

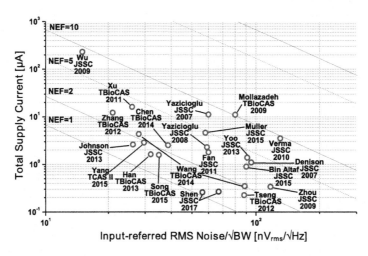

Figure 1.6 Noise efficiency factors of state-of-the-art instrumentation amplifiers for biopotential recording applications.

where $V_{rms,in}$ is the total input-referred noise, I_{tot} the total current drain, V_t the thermal voltage, and BW the -3-dB bandwidth of the system [39]. To minimize noise with a given current consumption or minimize current consumption with a upper-bound noise limit, various design techniques have been proposed and demonstrated to address these challenges [40–57,62,58–61]. State-of-the-art IAs typically have NEF of 2 to 10, and their measured NEF and performances are shown in Fig. 1.6. As a point of reference, measured NEF and performance of state-of Such techniques to minimize NEF include (i) utilizing the weak inversion region of CMOS operation to maximize transconductance efficiency [40,37,63,64], (ii) chopper stabilization techniques to reduce $1/f$ noise and other low-frequency noise [65,66,43,41,45,47,48], (iii) dynamic range manipulation to reduce power supply voltages [54,55] using spectrum–equalizing analog front-end [67,68], (iv) using current-reusing nMOS and pMOS input pairs to maximize transconductance and achieve a NEF below 2 [54,56,57,62,58].

Challenges in meeting the other specifications listed in Table 1.1 have also been addressed using various circuit techniques. For example, several DC-coupling IAs have been demonstrated in order to avoid external AC-coupling capacitors at the input of the AFE [42,43]. In these designs, electrode offsets are canceled by feedback currents via a DC-servo loop [42,43] or by capacitive feedback [69].

Integration of higher channel count on a single chip has been pursued, as well. Thus far, chips with approximately 100 to 300 data acquisition channels have been reported [70–72,54,73]. One of the strategies to reduce area and power consumption in order to maximize channel density is the use of scaled processes such as 65 nm CMOS [69], achieving 64 channels with a silicon area of 0.025 mm^2 per channel. For higher density, in some designs, an SAR ADC is shared by about 8 to 16

AFEs using a time-multiplexer [54]. In doing so, power efficient multiplexers [74] and time-interleaving sample-and-hold circuits in SAR ADCs have been demonstrated. Alternatively, a dedicated ADC per AFE channel has been also pursued due to its ease of integration with larger number of channels [44,69].

1.5.2 Integrated circuit interfaces for stimulation

Historically, electrical stimulation on the cortical surface was pioneered by Wilder Penfield [75] as intraoperative planning for epileptic patients, demonstrating the localized function of different regions of the cortex [76]. Since then, functional neural stimulation has been extensively investigated and developed during the past decades, making great progress for various clinical applications such as deep brain stimulation, cochlear implants, cardiac pacemakers, bladder control implants, and retinal prostheses. Given that many epilepsy patients already require implantation of ECoG monitoring instrumentation, there is a great opportunity for closed-loop electrical control of seizure activity at much higher resolution and precision than transcranial electric [77] and transcranial magnetic stimulation [78,79]. These embedded stimulators would not require any additional invasive risks, and could potentially prevent more drastic treatments such as partial removal of the cortex. An implantable recording and stimulation system can contain a digital signal processor capable of deciding when to stimulate [80]. Other applications of cortical stimulation include closed-loop brain computer interfaces (BCI) which aim to generate functional maps of the brain [81], restore somatosensory feedback [79], restore motor control to tetraplegics [82], aid stroke survivors [83,84], restore vision [85], reduce pain [76], or even change emotional state [86].

Pushing the form factor and channel density of the neural interface systems to the limit requires addressing several challenges in ASICs for stimulation. Smaller form factor and higher channel density require smaller electrode size, which limits charge transfer capacity for effective stimulation. Hence, higher voltage rails of more than ± 10 V and/or high-voltage processes are required typically [87–89], in turn this results in higher power consumption, larger silicon area, and system complexity to generate and handle high-voltage signals. Instead of maintaining a constant high-voltage power supply, some designs save power by generating a large power rail only when actively stimulating [80,90,91].

For further power savings, adiabatic stimulation has been also actively investigated. Adiabatic stimulators generate ramping power rails that closely follow the voltages at the stimulation electrode, minimizing unnecessary voltage drops across the current source employed for conventional constant-current stimulation. Various designs have been implemented with external capacitors [92], external inductors [93], and charge pumps [90]. Still, there is much room for improvement in the implementation of adiabatic stimulators in fully-integrated, miniaturized implantable ICs.

It is generally desired to minimize the area occupied per stimulation channel for high-density integration. To date, integration of 100 to 1600 channels has been achieved [87,94,95,88,89,96]. In order to integrate such high channel counts, programmability of waveform parameters, individual connectivity to each channel, and/or charge balancing need to be compromised to some extent. For example, groups of 4 to 8 electrodes in [97,94,96,98] can share a single digital-to-analog converter for optimized, high density integration.

Safety is of the utmost importance in chronic neural interfaces, so charge balancing is imperative [99]. Residual DC currents result in tissue damage, production of toxic byproducts, and electrode degradation [99]. However, it is quite challenging to assure charge balance for each channel in high-channel neural interface systems. One of the most straightforward strategies is to employ serial DC-blocking capacitors, inherently forcing the net DC current to be zero all the time. This method has been employed for neural stimulation applications [100–102] due to its intrinsic safety when area permits. However, the required blocking capacitance is often prohibitively large for on-chip integration, and is thus inadequate for high-density and/or miniaturized implants. Instead of external capacitors, capacitive electrodes made with high-k dielectric coatings have been investigated for safe neural interfaces [103,104,91]. Several other techniques for better charge balancing have been demonstrated: (i) shorting electrodes to ground [105], (ii) utilizing a discharging resistor [97], active current balancing by feedback control [106,107], generating additional balancing current pulses by monitoring electrode voltages [108], and embedded DAC calibration [109,96].

1.5.3 Integrated electrocortical online data processing

The integration of signal processing with neurophysiological sensing and actuation enables real-time online control strategies towards realizing adaptive, autonomous closed-loop systems for remediation of neurological disorders [110,111]. Online signal processing of ECoG data has tremendous potential to improve patient outcomes in diseases currently lacking therapy or requiring resection of otherwise healthy neural tissue such as intractable epilepsy. As one of the treatments for epilepsy, functional neurostimulation in response to detected seizures has been proved effective in reduction of seizures [112,113]. For real-time close-loop therapeutics, online automated seizure prediction and/or detection based on ECoG or EEG recordings of epileptic patients is imperative [114–117], and their on-chip implementation has been actively investigated and demonstrated utilizing extraction and classification of various signal features such as power spectral densities and wavelet coefficients [47,118,52,80,119,120].

In addition, ECoG has proven a powerful modality for BCI applications owing to richer features present in the higher resolution ECoG signals compared to surface EEG, which can be harnessed to more precisely infer sensory recognition, cognition, and motor function. Since ECoG-based BCI systems widely utilize spectral power density for their inputs [121], frequency band power extraction techniques have been implemented immediately following AFEs avoiding digitization and RF data transmission of whole ECoG raw signals [122,123].

Such on-chip real-time ECoG data processing offers two distinct advantages over offline as well as online off-chip processing. First, constrains on data bandwidth and power consumption on the implant can be largely relieved. In many implementations, raw recorded data is wirelessly streamed out and delivered to either a unit worn on the top of the head, or directly to a local base station such as a smartphone. The power of such approaches is typically proportional to the communication distance. Thus, the overall power consumption of designs that stream over long distances can be dominated by the power of communication circuits [47]. In order to reduce system-level power consumption, several on-chip data processing techniques have been applied for EEG- and ECoG-based BCI systems and epileptic seizure detection. By doing so, power consumption of RF data transmission can be drastically reduced [47]. Second, local processing may alleviate stringent latency and buffer memory requirements in the uplink transmission of data for external processing, especially where multiple implants are time-multiplexed between a common base station.

1.6. System considerations

1.6.1 Powering

Major challenges in implantable medical devices (IMD) for high-density brain activity monitoring are fundamentally posed by their target location. Some of these IMDs can be wholly placed on the cortex within a very limited geometry as shown in Fig. 1.3(A) In other cases, only the electrode array is placed on the cortex while the other components can be located in the empty space created by a craniotomy [124], or under the scalp with lead wires connected [125,126]. Regardless of placement, this constrained environment poses a difficult power challenge.

There are three primary methods for powering an implanted device: employing a battery, harvesting energy from the environment, and delivering power transcutaneously via a wireless power transmitter [127,128]. A natural first choice would be a battery, as they have been extensively used in other implantable applications such as pacemakers. While it makes sense to use a battery in a pacing application, where the power of the load circuit is small (microwatts) and a large physical volume is available such that the battery can last 10 years of more, the power consumption in high-density neural recording and stimulation applications is typically much larger (milliwatts), and the physical volume available for a large battery is small, combining to dramatically reducing operational lifetime prior to necessary surgical re-implantation. The medical risks of regular brain surgery and recovery, just to replace a battery, are unacceptable to most patients, and thus batteries are typically only employed in high-density neural applications as temporary energy storage in systems with a different power source: either energy scavenging or wireless power transfer.

Harvesting energy from ambient sources in the local environment has been a potentially attracting powering option since at least the 1970s during the development of cochlear implants. Many scavenging methods continue to be actively developed: (1) solar cells, (2) biofuel cells, (3) thermoelectric generators, (4) piezoelectric generators, (5) ambient RF, etc. While such approaches are theoretically attractive, the limited volume available near the brain, coupled with the stochastic nature of many energy harvesting sources, results in power that is too small and too variable to reliably operate multi-channel neural technologies.

The most popular means to power an implanted device with higher power than single-channel pacing applications is to wirelessly delivery power via a transcutaneous link. Power can be delivered transcutaneously using one of three primary mechanisms: (i) optics (typically near infrared light), (ii) acoustics (typically at ultrasound frequencies), and (iii) electromagnetics (either near-, mid-, or far-field waves). Each method can deliver from 10 µW up to the mW range of power. However, the total deliverable power highly depends on the geometry and make-up of the receiving transducer, along with the implant depth and orientation.

Optical powering through transmission of infrared light has a very short penetration depth of a few millimeters, limiting its utility to subcutaneous and very shallow implant applications [129–131]. Ultrasound, on the other hand, can penetrate much deeper into tissue, potentially powering implants located on the cortical surface. In fact, it has been demonstrated that ultrasound can more efficiently power mm-scale devices implanted deep into soft tissues than electromagnetic approaches [132]. However, it has also been shown that ultrasonic energy does not efficiently penetrate bone, limiting opportunities to directly power cortical implants from outside the skull. To overcome this, researchers have proposed two-tiered systems, where electromagnetic energy is coupled through the skull, then converted to acoustic energy via an intermediate transducer system, and finally delivered to the miniaturized implant through soft tissue [133]. However, in addition to nontrivial packaging and transducer design challenges, this is likely only a reasonable approach when the implant to be powered is either very deep, or very small (sub-mm scale). For these reasons, ultrasonic power delivery is not typically considered for ECoG systems.

The most popular transcutaneous power delivery approach utilizes electromagnetics. For devices implanted to a depth of a few centimeters, and that are on the order of mm-to-cm in diameter, near- or mid-field electromagnetic power transfer is generally considered to be the most efficient and practical method to power such devices. Near-field power transfer, which operates at frequencies up to approximately 100 MHz for typical implants, has been extensively used for cochlear implants [134], retinal prostheses [96,89], and various research IMD systems [135,124,136–138], and has been

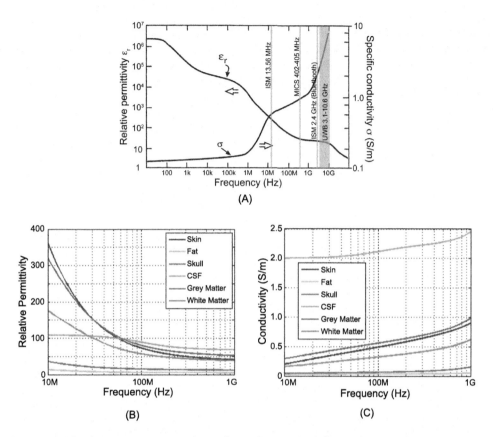

Figure 1.7 (A) Relative permittivity and specific conductivity over frequency range from 10 to 100 GHz with most popular frequency communication bands for ECoG implants [145,146]. (B) Relative permittivity and (C) specific conductivity of various kinds of tissues including skin, fat, skull, CSF, grey and white matter [147].

investigated and characterized to maximize its usage and power transfer efficiency for implants [139–144].

Most conventional designs operate in the near-field between 1 and 20 MHz, since it is well known that conductivity (and hence losses) in tissue increase at higher frequencies, as shown in Fig. 1.7. Operating at higher frequencies, it was previously argued, would encounter higher losses and thus be less efficient. In addition, governmental regulatory agencies limit the amount of power that can be dissipated in tissue for safety reasons – the US Federal Communications Commission (FCC) sets a specific absorption rate (SAR) of less than 1.6 W/kg, for example. For these reasons, conventional transcutaneous power transfer links operate in the low-MHz range, often at the 6.78 and 13.56 MHz ISM bands [143,144,127].

However, it is also well known that the quality factor and radiation resistance of electrically-small coil antennas increases with increasing frequency. Thus, miniaturized implants, which have electrically small coils for wireless power reception, tend to prefer to operate at higher frequencies, at least in air. In biological tissues, the trade-off between coil design and tissue losses results in an optimal frequency for wireless power transmission where efficiency is maximized. For example, the inductance of coils located on miniaturized, mm-scale implants ranges from 10 to 100 nH [69,91,148,149]. To compensate reduced magnetic flux through the miniaturized receiving coil, the carrier frequency for wireless power transfer should be increased, often into the hundreds of MHz to single-digit GHz range [150,151,149,148,69]. These prior studies have demonstrated that it is possible to efficiently deliver milliwatts of power to small, implanted devices under regulatory limits, and thus electromagnetic approaches are the primary means to deliver power to implanted ECoG devices.

1.6.2 Wireless data communication

Implanted ECoG monitoring devices need to convey the acquired data to the external world through wireless communication. The information received by the external base station can be monitored, processed and used by users and care takers for health monitoring, treatments, or scientific research. For ECoG monitoring implants, data transmission from the implant to the external device, known as uplink or backward telemetry, requires much higher data rate and is subject to more stringent power consumption constraints than data transmission from the external device to the implant, known as downlink or forward telemetry, because the available power and geometric volume are much smaller on the implanted side than the external side. This power constraint on backward data transmission is more exacerbated as the number of channels increases, as higher data rates are required in a typically more compact area, leading to severe power density challenges.

Backward data communications typically employ electromagnetics operating either in the far- or near-field. Far-field communication uses electromagnetic radiation to transmit data over a distance much longer than the size of the actual device. Hence, the implant can send data to a external base station located up to a few meters, such as mobile phone. Far-field up-conversion transmitters are currently the most well-established communication technology. Due to the wide availability of far-field radio products and a myriad of different infrastructures (e.g., Bluetooth Low Energy, WiFi, etc.), far-field radios can be quickly adopted for robust operation [124]. However, even state-of-the-art low-power radios consume >1 nJ/bit [127,152], which is order-of-magnitude larger than what typical ECoG recording IMDs require.

As an alternative far-field transmission method, impulse radio ultra-wide band (IR-UWB) transmission has emerged recently due to its low power consumption in the range of a few tens of pJ/bit [153–155]. Avoiding generation of a carrier with an accurate frequency, non–coherent IR–UWB transmitters generates short pulses with on–off keying (OOK) or pulse position modulation (PPM). Due to its high and wide frequency range (3.1–10 GHz), data rates of more than 10 Mbps with tens of pJ/bit have been reported [156]. In addition, antennas for this type of transmission do not need to be large. However, there are a couple of critical reasons against their usage for IMDs [127]. Foremost, their peak transmission power is large due to the inherently duty-cycled nature of IR–UWB transmitters. Thus, while the average power may be low, a large high-quality power supply with a large battery of capacitor is required to supply large peak currents, which may be prohibitively large for many ECoG applications. Moreover, since IR–UWB operates at very high frequency over 3 GHz, tissue absorption rate is higher.

In contrast to far-field communication methods, near-field radios operate over short distances, typically within one wavelength of the carrier frequency, and are thus suitable for use when an external device is located directly on the head. In fact, since this configuration is naturally present in wirelessly-powered devices, near-field communication can easily be implemented along with this wireless powering. One of the most popular data communication methods that can be implemented along with forward power delivery is the backscattering method [157–161]. This method modulates the load conditions of forward powering signals, and reflecting this energy back to the interrogator. Since only a single switch needs to be driven, the power consumed on the implant side is minimal, as carrier generation and active driving of an antenna are not required. Since, with this technology, a few pJ/bit to tens of pJ/bit can be achieved [161], it has been widely adopted by various IMDs [162,89,163,69,164–166]. However, external data reception in backscattering systems can be challenging in some cases because of the large power difference between the large power carrier signal and the weak backscattered signal.

Forward telemetry for neural recording IMDs is typically used for sending configuration bits to the implant, requiring a relatively low data rate typically much less than 100 kbps. Hence, amplitude shift keying (ASK) has been widely employed in such IMDs [167,168,91], modulating the power carrier signals. For IMDs with stimulation capability, forward telemetry for closed-loop operation is typically time-multiplexed with backward telemetry.

1.6.3 Hermetic encapsulation

Implanted devices containing silicon ICs need packaging in a protective enclosure to mitigate corrosion and other contamination by surrounding electrolyte in the body [169,170]. Thus far, titanium-, glass- or ceramic-based enclosures have been the primary means for hermetic sealing in long-term implants, since these hard materials have been shown to be biocompatible and impermeable to water [169]. Even though such hard enclosures are used in the majority of long-term implants [171–174,114,175, 176,124,3,177,178], their very large volume and weight, typically much larger than the ICs and supporting components they contain, prohibits their use in high-dimensional neural interfaces heavily constrained by anatomical space such as retinal prostheses [179] and μECoG arrays [30]. In addition, hard packaging requires intricate methods for hermetic sealing of feedthroughs to polymer insulated extensions of the implant such as electrode array cabling, limiting the density of electrode channels due to feedthrough channel spacing requirements.

To overcome these challenges, conformal coating of integrated electronics with polymers such as polyimide, silicone, and parylene-C have been investigated as alternatives. They are superior over metal, glass, and ceramic hard seals in miniaturization, flexibility, and compatibility with the semiconductor process [180]. However, polymers are susceptible to degradation and are not long-term impermeable to body fluids. While polymer encapsulation has been used for relatively simple and short-term (less than 1–2 years) implants, substantial improvements are needed for viable solutions to long-term hermetic encapsulation [170,181]. Recent next-generation advances in miniaturized hermetic sealing of silicon integrated circuits, such as multi-layer multi-material coating [182], and encapsulation using liquid crystal polymers (LCP) [180], are promising developments towards highly miniaturized implantable electronics for chronic clinical use. Furthermore, recent advances in dissolvable flexible electronics [183,184] offer alternatives to hermetic encapsulation for acute applications without the need for post-use surgical extraction.

1.7. Conclusion

In this chapter we highlighted the importance of high density electrocorticography for brain activity mapping, brain computer interfaces, and treatments for neurological disorders. We briefly reviewed the critical design challenges on ECoG interface systems in their major aspects including recording and stimulation circuitry, wireless power and data communications, communication and hermetic encapsulation, each of which will be discussed more in detail in the following chapters.

References

[1] The White House, The BRAIN initiative, [Online]. Available: https://www.whitehouse.gov/BRAIN.
[2] The National Institute of Health, BRAIN working group report: BRAIN 2025 – a scientific vision, [Online]. Available: http://braininitiative.nih.gov/2025/BRAIN2025.pdf.
[3] M. Hirata, K. Matsushita, T. Suzuki, T. Yoshida, F. Sato, S. Morris, T. Yanagisawa, T. Goto, M. Kawato, T. Yoshimine, A fully-implantable wireless system for human brain–machine interfaces using brain surface electrodes: W-HERBS, IEICE Transactions on Communications E94b (9) (2011) 2448–2453.
[4] P.L. Nunez, Electric and magnetic fields produced by the brain, in: J. Wolpaw, E.W. Wolpaw (Eds.), Brain–Computer Interfaces: Principles and Practice, Oxford University Press, 2012, pp. 171–212.
[5] R.B. Reilly, Neurology: central nervous system, in: J.G. Webster (Ed.), The Physiological Measurement Handbook, CRC Press, New York, 2014, pp. 171–212.

[6] M. Fukushima, Z.C. Chao, N. Fujii, Studying brain functions with mesoscopic measurements: advances in electrocorticography for non-human primates, Current Opinion in Neurobiology 32 (2015) 124–131.

[7] H. Lee, R.V. Bellamkonda, W. Sun, M.E. Levenston, Biomechanical analysis of silicon microelectrode-induced strain in the brain, Journal of Neural Engineering 2 (4) (2005) 81.

[8] G.C. McConnell, H.D. Rees, A.I. Levey, C.-A. Gutekunst, R.E. Gross, R.V. Bellamkonda, Implanted neural electrodes cause chronic, local inflammation that is correlated with local neurodegeneration, Journal of Neural Engineering 6 (5) (2009) 056003.

[9] L. Karumbaiah, T. Saxena, D. Carlson, K. Patil, R. Patkar, E.A. Gaupp, M. Betancur, G.B. Stanley, L. Carin, R.V. Bellamkonda, Relationship between intracortical electrode design and chronic recording function, Biomaterials 34 (33) (2013) 8061–8074.

[10] V.S. Polikov, P.A. Tresco, W.M. Reichert, Response of brain tissue to chronically implanted neural electrodes, Journal of Neuroscience Methods 148 (1) (2005) 1–18.

[11] W.J. Freeman, L.J. Rogers, M.D. Holmes, D.L. Silbergeld, Spatial spectral analysis of human electrocorticograms including the alpha and gamma bands, Journal of Neuroscience Methods 95 (2) (2000) 111–121.

[12] E.C. Leuthardt, Z. Freudenberg, D. Bundy, J. Roland, Microscale recording from human motor cortex: implications for minimally invasive electrocorticographic brain–computer interfaces, Neurosurgical Focus 27 (1) (2009) E10.

[13] M.W. Slutzky, L.R. Jordan, T. Krieg, M. Chen, D.J. Mogul, L.E. Miller, Optimal spacing of surface electrode arrays for brain–machine interface applications, Journal of Neural Engineering 7 (2) (2010) 026004.

[14] R.J. Staba, C.L. Wilson, A. Bragin, I. Fried, J. Engel, Quantitative analysis of high-frequency oscillations (80–500 Hz) recorded in human epileptic hippocampus and entorhinal cortex, Journal of Neurophysiology 88 (4) (2002) 1743–1752.

[15] W.J. Freeman, M.D. Holmes, B.C. Burke, S. Vanhatalo, Spatial spectra of scalp EEG and EMG from awake humans, Clinical Neurophysiology 114 (6) (2003) 1053–1068.

[16] T. Ball, M. Kern, I. Mutschler, A. Aertsen, A. Schulze-Bonhage, Signal quality of simultaneously recorded invasive and non-invasive EEG, NeuroImage 46 (3) (2009) 708–716.

[17] P.L. Nunez, R. Srinivasan, Fallacies in EEG, in: Electric Fields of the Brain: The Neurophysics of EEG, 2nd ed., Oxford University Press, New York, NY, 2006, pp. 56–98, ch. 2.

[18] H. Berger, Electroencephalogram in humans, Archiv für Psychiatrie und Nervenkrankheiten 87 (1929) 527–570.

[19] P.L. Nunez, R. Srinivasan, Electric fields and currents in biological tissue, in: Electric Fields of the Brain: The Neurophysics of EEG, 2nd ed., Oxford University Press, New York, NY, 2006, pp. 147–202, ch. 4.

[20] E.C. Leuthardt, G. Schalk, J.R. Wolpaw, J.G. Ojemann, D.W. Moran, A brain–computer interface using electrocorticographic signals in humans, Journal of Neural Engineering 1 (2) (2004) 63.

[21] S. Kellis, B. Greger, S. Hanrahan, P. House, R. Brown, Platinum microwire for subdural electrocorticography over human neocortex: millimeter-scale spatiotemporal dynamics, in: Proceedings of the Annual International Conference of the IEEE Engineering in Medicine and Biology Society, 2011, pp. 4761–4765.

[22] S. Kellis, S. Hanrahan, T. Davis, P.A. House, R. Brown, B. Greger, Decoding hand trajectories from micro-electrocorticography in human patients, in: Proceedings of the Annual International Conference of the IEEE Engineering in Medicine and Biology Society, 2012, pp. 4091–4094.

[23] S. Morris, M. Hirata, H. Sugata, T. Goto, K. Matsushita, T. Yanagisawa, Y. Saitoh, H. Kishima, T. Yoshimine, Patient-specific cortical electrodes for sulcal and gyral implantation, IEEE Transactions on Biomedical Engineering 62 (4) (2015) 1034–1041.

[24] B. Rubehn, C. Bosman, R. Oostenveld, P. Fries, T. Stieglitz, A MEMS-based flexible multichannel ECoG-electrode array, Journal of Neural Engineering 6 (3) (2009) 036003.

[25] E. Tolstosheeva, V. Gordillo-González, V. Biefeld, L. Kempen, S. Mandon, A.K. Kreiter, W. Lang, A multi-channel, flex-rigid ECoG microelectrode array for visual cortical interfacing, Sensors 15 (1) (2015) 832–854.

[26] F. Kohler, M. Schuettler, T. Stieglitz, Parylene-coated metal tracks for neural electrode arrays – fabrication approaches and improvements utilizing different laser systems, in: Proceedings of the Annual International Conference of the IEEE Engineering in Medicine and Biology Society, 2012, pp. 5130–5133.

[27] T.J. Richner, S. Thongpang, S.K. Brodnick, A.A. Schendel, R.W. Falk, L.A. Krugner-Higby, R. Pashaie, J.C. Williams, Optogenetic micro-electrocorticography for modulating and localizing cerebral cortex activity, Journal of Neural Engineering 11 (1) (2014) 016010.

[28] E. Castagnola, L. Maiolo, E. Maggiolini, A. Minotti, M. Marrani, F. Maita, A. Pecora, G.N. Angotzi, A. Ansaldo, M. Boffini, L. Fadiga, G. Fortunato, D. Ricci, PEDOT-CNT-coated low-impedance, ultra-flexible, and brain-conformable micro-ECoG arrays, IEEE Transactions on Neural Systems and Rehabilitation Engineering 23 (3) (2015) 342–350.

[29] D.H. Kim, J. Viventi, J.J. Amsden, J.L. Xiao, L. Vigeland, Y.S. Kim, J.A. Blanco, B. Panilaitis, E.S. Frechette, D. Contreras, D.L. Kaplan, F.G. Omenetto, Y.G. Huang, K.C. Hwang, M.R. Zakin, B. Litt, J.A. Rogers, Dissolvable films of silk fibroin for ultrathin conformal bio-integrated electronics, Nature Materials 9 (6) (2010) 511–517.

[30] J. Viventi, D.H. Kim, L. Vigeland, E.S. Frechette, J.A. Blanco, Y.S. Kim, A.E. Avrin, V.R. Tiruvadi, S.W. Hwang, A.C. Vanleer, D.F. Wulsin, K. Davis, C.E. Gelber, L. Palmer, J. Van der Spiegel, J. Wu, J.L. Xiao, Y.G. Huang, D. Contreras, J.A. Rogers, B. Litt, Flexible, foldable, actively multiplexed, high-density electrode array for mapping brain activity in vivo, Nature Neuroscience 14 (12) (2011) 1599–U138.

[31] K.Y. Kwon, B. Sirowatka, A. Weber, W. Li, Opto-ECoG array: a hybrid neural interface with transparent ECoG electrode array and integrated LEDs for optogenetics, IEEE Transactions on Biomedical Circuits and Systems 7 (5) (2013) 593–600.

[32] S. Ha, C. Kim, Y.M. Chi, A. Akinin, C. Maier, A. Ueno, G. Cauwenberghs, Integrated circuits and electrode interfaces for noninvasive physiological monitoring, IEEE Transactions on Biomedical Engineering 61 (5) (2014) 1522–1537.

[33] R. Grech, T. Cassar, J. Muscat, K.P. Camilleri, S.G. Fabri, M. Zervakis, P. Xanthopoulos, V. Sakkalis, B. Vanrumste, Review on solving the inverse problem in EEG source analysis, Journal of Neuroengineering and Rehabilitation 5 (2008).

[34] Y.-C. Chen, H.-L. Hsu, Y.-T. Lee, H.-C. Su, S.-J. Yen, C.-H. Chen, W.-L. Hsu, T.-R. Yew, S.-R. Yeh, D.-J. Yao, Y.-C. Chang, H. Chen, An active, flexible carbon nanotube microelectrode array for recording electrocorticograms, Journal of Neural Engineering 8 (3) (2011) 034001.

[35] A.A. Schendel, K.W. Eliceiri, J.C. Williams, Advanced materials for neural surface electrodes, Current Opinion in Solid State and Materials Science 18 (6) (2014) 301–307.

[36] D. Khodagholy, T. Doublet, P. Quilichini, M. Gurfinkel, P. Leleux, A. Ghestem, E. Ismailova, T. Hervé, S. Sanaur, C. Bernard, G.G. Malliaras, In vivo recordings of brain activity using organic transistors, Nature Communications 4 (2013) 1575.

[37] R.R. Harrison, The design of integrated circuits to observe brain activity, Proceedings of the IEEE 96 (7) (2008) 1203–1216.

[38] X. Tong, M. Ghovanloo, Multichannel wireless neural recording AFE architectures analysis, modeling, and tradeoffs, IEEE Design & Test 33 (4) (2016) 24–36.

[39] M.S.J. Steyaert, W.M.C. Sansen, Z.Y. Chang, A micropower low-noise monolithic instrumentation amplifier for medical purposes, IEEE Journal of Solid-State Circuits 22 (6) (1987) 1163–1168.

[40] R.R. Harrison, C. Charles, A low-power low-noise CMOS amplifier for neural recording applications, IEEE Journal of Solid-State Circuits 38 (6) (2003) 958–965.

[41] T. Denison, K. Consoer, W. Santa, A.T. Avestruz, J. Cooley, A. Kelly, A 2 μW 100 nV/rtHz chopper-stabilized instrumentation amplifier for chronic measurement of neural field potentials, IEEE Journal of Solid-State Circuits 42 (12) (2007) 2934–2945.

[42] R.F. Yazicioglu, P. Merken, R. Puers, C. Van Hoof, A 60 μW 60 nV/rtHz readout front-end for portable biopotential acquisition systems, IEEE Journal of Solid-State Circuits 42 (5) (2007) 1100–1110.

[43] R.F. Yazicioglu, P. Merken, R. Puers, C. van Hoof, A 200 μW eight-channel EEG acquisition ASIC for ambulatory EEG systems, IEEE Journal of Solid-State Circuits 43 (12) (2008) 3025–3038.

[44] M. Mollazadeh, K. Murari, G. Cauwenberghs, N. Thakor, Micropower CMOS integrated low-noise amplification, filtering, and digitization of multimodal neuropotentials, IEEE Transactions on Biomedical Circuits and Systems 3 (1) (2009) 1–10.

[45] R. Wu, K.A.A. Makinwa, J.H. Huijsing, A chopper current-feedback instrumentation amplifier with a 1 mHz 1/f noise corner and an AC-coupled ripple reduction loop, IEEE Journal of Solid-State Circuits 44 (12) (2009) 3232–3243.

[46] X. Zou, X. Xu, L. Yao, Y. Lian, A 1-V 450-nW fully integrated programmable biomedical sensor interface chip, IEEE Journal of Solid-State Circuits 44 (4) (2009) 1067–1077.

[47] N. Verma, A. Shoeb, J. Bohorquez, J. Dawson, J. Guttag, A.P. Chandrakasan, A micro-power EEG acquisition SoC with integrated feature extraction processor for a chronic seizure detection system, IEEE Journal of Solid-State Circuits 45 (4) (2010) 804–816.

[48] J. Xu, R.F. Yazicioglu, B. Grundlehner, P. Harpe, K.A.A. Makinwa, C. Van Hoof, A 160 μW 8-channel active electrode system for EEG monitoring, IEEE Transactions on Biomedical Circuits and Systems 5 (6) (2011) 555–567.

[49] Q.W. Fan, F. Sebastiano, J.H. Huijsing, K.A.A. Makinwa, A 1.8 μW 60 nV/rtHz capacitively-coupled chopper instrumentation amplifier in 65 nm CMOS for wireless sensor nodes, IEEE Journal of Solid-State Circuits 46 (7) (2011) 1534–1543.

[50] F. Zhang, J. Holleman, B.P. Otis, Design of ultra-low power biopotential amplifiers for biosignal acquisition applications, IEEE Transactions on Biomedical Circuits and Systems 6 (4) (2012) 344–355.

[51] Y. Tseng, Y.C. Ho, S.T. Kao, C.C. Su, A 0.09 μW low power front-end biopotential amplifier for biosignal recording, IEEE Transactions on Biomedical Circuits and Systems 6 (5) (2012) 508–516.

[52] J. Yoo, L. Yan, D. El-Damak, M.A. Bin Altaf, A.H. Shoeb, A.P. Chandrakasan, An 8-channel scalable EEG acquisition SoC with patient-specific seizure classification and recording processor, IEEE Journal of Solid-State Circuits 48 (1) (2013) 214–228.

[53] B. Johnson, A. Molnar, An orthogonal current-reuse amplifier for multi-channel sensing, IEEE Journal of Solid-State Circuits 48 (6) (2013) 1487–1496.

[54] D. Han, Y. Zheng, R. Rajkumar, G.S. Dawe, M. Je, A 0.45 V 100-channel neural-recording IC with sub-μW/channel consumption in 0.18 μm CMOS, IEEE Transactions on Biomedical Circuits and Systems 7 (6) (2013) 735–746.

[55] Y. Chen, A. Basu, L. Liu, X. Zou, R. Rajkumar, G.S. Dawe, M. Je, A digitally assisted, signal folding neural recording amplifier, IEEE Transactions on Biomedical Circuits and Systems 8 (4) (2014) 528–542.

[56] T.-Y. Wang, M.-R. Lai, C.M. Twigg, S.-Y. Peng, A fully reconfigurable low-noise biopotential sensing amplifier with 1.96 noise efficiency factor, IEEE Transactions on Biomedical Circuits and Systems 8 (3) (2014) 411–422.

[57] S. Song, M. Rooijakkers, P. Harpe, C. Rabotti, M. Mischi, A.H.M. van Roermund, E. Cantatore, A low-voltage chopper-stabilized amplifier for fetal ECG monitoring with a 1.41 power efficiency factor, IEEE Transactions on Biomedical Circuits and Systems 9 (2) (2015) 237–247.

[58] P. Harpe, H. Gao, R. van Dommele, E. Cantatore, A. van Roermund, A 0.20mm^2 3 nW signal acquisition IC for miniature sensor nodes in 65 nm CMOS, IEEE Journal of Solid-State Circuits 51 (1) (2016) 240–248.

[59] F.M. Yaul, A.P. Chandrakasan, A noise-efficient 36 nV/ $\sqrt{}$ Hz chopper amplifier using an inverter-based 0.2-V supply input stage, IEEE Journal of Solid-State Circuits 52 (11) (2017) 3032–3042.

[60] L. Shen, N. Lu, N. Sun, A 1-V 0.25-μW inverter stacking amplifier with 1.07 noise efficiency factor, IEEE Journal of Solid-State Circuits 53 (3) (2018) 896–905.

[61] J. Zhang, H. Zhang, Q. Sun, R. Zhang, A low-noise, low-power amplifier with current-reused OTA for ECG recordings, IEEE Transactions on Biomedical Circuits and Systems 12 (3) (2018) 700–708.

[62] T. Yang, J. Holleman, An ultralow-power low-noise CMOS biopotential amplifier for neural recording, IEEE Transactions on Circuits and Systems II: Express Briefs 62 (10) (2015) 927–931.

[63] R. Sarpeshkar, Ultra Low Power Bioelectronics: Fundamentals, Biomedical Applications, and Bio-Inspired Systems, Cambridge University Press, 2010.

[64] S. Ha, C. Kim, Y.M. Chi, G. Cauwenberghs, Low-power integrated circuit design for wearable biopotential sensing, in: E. Sazonov, M.R. Neuman (Eds.), Wearable Sensors, Academic Press, Oxford, 2014, pp. 323–352.

[65] C.C. Enz, G.C. Temes, Circuit techniques for reducing the effects of op-amp imperfections: autozeroing, correlated double sampling, and chopper stabilization, Proceedings of the IEEE 84 (11) (1996) 1584–1614.

[66] C. Menolfi, Q.T. Huang, A low-noise CMOS instrumentation amplifier for thermoelectric infrared detectors, IEEE Journal of Solid-State Circuits 32 (7) (1997) 968–976.

[67] W. Smith, B. Mogen, E. Fetz, B. Otis, A spectrum-equalizing analog front end for low-power electrocorticography recording, in: Proceedings of the IEEE European Solid State Circuits Conference, 2014, pp. 107–110.

[68] S.-Y. Park, J. Cho, K. Na, E. Yoon, Toward 1024-channel parallel neural recording: modular $\Delta - \Delta\Sigma$ analog front-end architecture with 4.84fJ/c-s·mm^2 energy-area product, in: Symposium on VLSI Circuits Digest of Technical Papers, 2015, pp. C112–C113.

[69] R. Muller, H.-P. Le, W. Li, P. Ledochowitsch, S. Gambini, T. Bjorninen, A. Koralek, J.M. Carmena, M.M. Maharbiz, E. Alon, J.M. Rabaey, A minimally invasive 64-channel wireless μECoG implant, IEEE Journal of Solid-State Circuits 50 (1) (2015) 344–359.

[70] R.R. Harrison, P.T. Watkins, R.J. Kier, R.O. Lovejoy, D.J. Black, B. Greger, F. Solzbacher, A low-power integrated circuit for a wireless 100-electrode neural recording system, IEEE Journal of Solid-State Circuits 42 (1) (2007) 123–133.

[71] J.N.Y. Aziz, K. Abdelhalim, R. Shulyzki, R. Genov, B.L. Bardakjian, M. Derchansky, D. Serletis, P.L. Carlen, 256-channel neural recording and delta compression microsystem with 3D electrodes, IEEE Journal of Solid-State Circuits 44 (3) (2009) 995–1005.

[72] H. Gao, R.M. Walker, P. Nuyujukian, K.A.A. Makinwa, K.V. Shenoy, B. Murmann, T.H. Meng, HermesE: a 96-channel full data rate direct neural interface in 0.13 μm CMOS, IEEE Journal of Solid-State Circuits 47 (4) (2012) 1043–1055.

[73] R. Shulyzki, K. Abdelhalim, A. Bagheri, M.T. Salam, C.M. Florez, J.L. Perez Velazquez, P.L. Carlen, R. Genov, 320-Channel active probe for high-resolution neuromonitoring and responsive neurostimulation, IEEE Transactions on Biomedical Circuits and Systems 9 (1) (2015) 34–49.

[74] L. Yan, P. Harpe, V.R. Pamula, M. Osawa, Y. Harada, K. Tamiya, C. Van Hoof, R.F. Yazicioglu, A 680 nA ECG acquisition IC for leadless pacemaker applications, IEEE Transactions on Biomedical Circuits and Systems 8 (6) (2014) 779–786.

[75] W. Penfield, E. Boldrey, Somatic motor and sensory representation in the cerebral cortex of man as studied by electrical stimulation, Brain 60 (4) (1937) 389–443.

[76] L. Mazzola, J. Isnard, R. Peyron, F. Mauguière, Stimulation of the human cortex and the experience of pain: Wilder Penfield's observations revisited, Brain 135 (2) (2012) 631–640.

[77] A. Berényi, M. Belluscio, D. Mao, G. Buzsáki, Closed-loop control of epilepsy by transcranial electrical stimulation, Science 337 (6095) (2012) 735–737.

[78] D.J. Mogul, W. van Drongelen, Electrical control of epilepsy, Annual Review of Biomedical Engineering 16 (2014) 483–504.

[79] L.A. Johnson, J.D. Wander, D. Sarma, D.K. Su, E.E. Fetz, J.G. Ojemann, Direct electrical stimulation of the somatosensory cortex in humans using electrocorticography electrodes: a qualitative and quantitative report, Journal of Neural Engineering 10 (3) (2013) 036021.

[80] W.-M. Chen, H. Chiueh, T.-J. Chen, C.-L. Ho, C. Jeng, M.-D. Ker, C.-Y. Lin, Y.-C. Huang, C.-W. Chou, T.-Y. Fan, M.-S. Cheng, Y.-L. Hsin, S.-F. Liang, Y.-L. Wang, F.-Z. Shaw, Y.-H. Huang, C.-H. Yang, C.-Y. Wu, A fully integrated 8-channel closed-loop neural-prosthetic CMOS SoC for real-time epileptic seizure control, IEEE Journal of Solid-State Circuits 49 (1) (2014) 232–247.

[81] S. Borchers, M. Himmelbach, N. Logothetis, H.O. Karnath, Direct electrical stimulation of human cortex – the gold standard for mapping brain functions?, Nature Reviews Neuroscience 13 (1) (2012) 63–70.

[82] B. Jarosiewicz, N.Y. Masse, D. Bacher, S.S. Cash, E. Eskandar, G. Friehs, J.P. Donoghue, L.R. Hochberg, Advantages of closed-loop calibration in intracortical brain–computer interfaces for people with tetraplegia, Journal of Neural Engineering 10 (4) (2013) 046012.

[83] R. Levy, S. Ruland, M. Weinand, D. Lowry, R. Dafer, R. Bakay, Cortical stimulation for the rehabilitation of patients with hemiparetic stroke: a multicenter feasibility study of safety and efficacy, Journal of Neurosurgery 108 (4) (2008) 707–714.

[84] M.A. Edwardson, T.H. Lucas, J.R. Carey, E.E. Fetz, New modalities of brain stimulation for stroke rehabilitation, Experimental Brain Research 224 (3) (2013) 335–358.

[85] P.M. Lewis, H.M. Ackland, A.J. Lowery, J.V. Rosenfeld, Restoration of vision in blind individuals using bionic devices: a review with a focus on cortical visual prostheses, Brain Research 1595 (2015) 51–73.

[86] F. Caruana, P. Avanzini, F. Gozzo, S. Francione, F. Cardinale, G. Rizzolatti, Mirth and laughter elicited by electrical stimulation of the human anterior cingulate cortex, Cortex 71 (2015) 323–331.

[87] M. Ortmanns, A. Rocke, M. Gehrke, H.J. Tiedtke, A 232-channel epiretinal stimulator ASIC, IEEE Journal of Solid-State Circuits 42 (12) (2007) 2946–2959.

[88] C. Kuanfu, Z. Yang, H. Linh, J. Weiland, M. Humayun, L. Wentai, An integrated 256-channel epiretinal prosthesis, IEEE Journal of Solid-State Circuits 45 (9) (2010) 1946–1956.

[89] Y.-K. Lo, K. Chen, P. Gad, W. Liu, A fully-integrated high-compliance voltage SoC for epi-retinal and neural prostheses, IEEE Transactions on Biomedical Circuits and Systems 7 (6) (2013) 761–772.

[90] W. Biederman, D.J. Yeager, N. Narevsky, J. Leverett, R. Neely, J.M. Carmena, E. Alon, J.M. Rabaey, A 4.78 mm^2 fully-integrated neuromodulation SoC combining 64 acquisition channels with digital compression and simultaneous dual stimulation, IEEE Journal of Solid-State Circuits 50 (4) (2015) 1038–1047.

[91] S. Ha, A. Akinin, J. Park, C. Kim, H. Wang, C. Maier, G. Cauwenberghs, P.P. Mercier, A 16-channel wireless neural interfacing SoC with RF-powered energy-replenishing adiabatic stimulation, in: Symposium on VLSI Circuits Digest of Technical Papers, 2015, pp. C106–C107.

[92] S.K. Kelly, J.L. Wyatt, A power-efficient neural tissue stimulator with energy recovery, IEEE Transactions on Biomedical Circuits and Systems 5 (1) (2011) 20–29.

[93] S.K. Arfin, R. Sarpeshkar, An energy-efficient, adiabatic electrode stimulator with inductive energy recycling and feedback current regulation, IEEE Transactions on Biomedical Circuits and Systems 6 (1) (2012) 1–14.

[94] A. Rothermel, V. Wieczorek, L. Liu, A. Stett, M. Gerhardt, A. Harscher, S. Kibbel, A 1600-pixel subretinal chip with DC-free terminals and ±2V supply optimized for long lifetime and high stimulation efficiency, in: IEEE International Solid-State Circuits Conference Digest of Technical Papers, 2008, pp. 144–145.

[95] B.K. Thurgood, D.J. Warren, N.M. Ledbetter, G.A. Clark, R.R. Harrison, A wireless integrated circuit for 100-channel charge-balanced neural stimulation, IEEE Transactions on Biomedical Circuits and Systems 3 (6) (2009) 405–414.

[96] M. Monge, M. Raj, M.H. Nazari, C. Han-Chieh, Z. Yu, J.D. Weiland, M.S. Humayun, T. Yu-Chong, A. Emami, A fully intraocular high-density self-calibrating epiretinal prosthesis, IEEE Transactions on Biomedical Circuits and Systems 7 (6) (2013) 747–760.

[97] M. Sivaprakasam, L. Wentai, M.S. Humayun, J.D. Weiland, A variable range bi-phasic current stimulus driver circuitry for an implantable retinal prosthetic device, IEEE Journal of Solid-State Circuits 40 (3) (2005) 763–771.

[98] M. Yip, J. Rui, H.H. Nakajima, K.M. Stankovic, A.P. Chandrakasan, A fully-implantable cochlear implant SoC with piezoelectric middle-ear sensor and arbitrary waveform neural stimulation, IEEE Journal of Solid-State Circuits 50 (1) (2015) 214–229.

[99] D.R. Merrill, M. Bikson, J.G.R. Jefferys, Electrical stimulation of excitable tissue: design of efficacious and safe protocols, Journal of Neuroscience Methods 141 (2) (2005) 171–198.

[100] C.Q. Huang, R.K. Shepherd, P.M. Carter, P.M. Seligman, B. Tabor, Electrical stimulation of the auditory nerve: direct current measurement in vivo, IEEE Transactions on Biomedical Engineering 46 (4) (1999) 461–470.

[101] G.J. Suaning, N.H. Lovell, CMOS neurostimulation ASIC with 100 channels, scaleable output, and bidirectional radio-frequency telemetry, IEEE Transactions on Biomedical Engineering 48 (2) (2001) 248–260.

[102] S.K. Arfin, M.A. Long, M.S. Fee, R. Sarpeshkar, Wireless neural stimulation in freely behaving small animals, Journal of Neurophysiology 102 (1) (2009) 598–605.

[103] I. Schoen, P. Fromherz, Extracellular stimulation of mammalian neurons through repetitive activation of Na^+ channels by weak capacitive currents on a silicon chip, Journal of Neurophysiology 100 (1) (2008) 346–357.

[104] A. Lambacher, V. Vitzthum, R. Zeitler, M. Eickenscheidt, B. Eversmann, R. Thewes, P. Fromherz, Identifying firing mammalian neurons in networks with high-resolution multi-transistor array (MTA), Applied Physics A: Materials Science & Processing 102 (1) (2011) 1–11.

[105] W. Liu, K. Vichienchom, M. Clements, S.C. DeMarco, C. Hughes, E. McGucken, M.S. Humayun, E. De Juan, J.D. Weiland, R. Greenberg, A neuro-stimulus chip with telemetry unit for retinal prosthetic device, IEEE Journal of Solid-State Circuits 35 (10) (2000) 1487–1497.

[106] J.-J. Sit, R. Sarpeshkar, A low-power blocking-capacitor-free charge-balanced electrode-stimulator chip with less than 6 nA DC error for 1-mA full-scale stimulation, IEEE Transactions on Biomedical Circuits and Systems 1 (3) (2007) 172–183.

[107] E. Noorsal, K. Sooksood, X. Hongcheng, R. Hornig, J. Becker, M. Ortmanns, A neural stimulator frontend with high-voltage compliance and programmable pulse shape for epiretinal implants, IEEE Journal of Solid-State Circuits 47 (1) (2012) 244–256.

[108] K. Sooksood, T. Stieglitz, M. Ortmanns, An active approach for charge balancing in functional electrical stimulation, IEEE Transactions on Biomedical Circuits and Systems 4 (3) (2010) 162–170.

[109] E. Greenwald, C. Cheng, N. Thakor, C. Maier, G. Cauwenberghs, A CMOS neurostimulator with on-chip DAC calibration and charge balancing, in: Proceedings of the IEEE Biomedical Circuits and Systems Conference, 2013, pp. 89–92.

[110] F.D. Broccard, T. Mullen, Y.M. Chi, D. Peterson, J.R. Iversen, M. Arnold, K. Kreutz-Delgado, T.-P. Jung, S. Makeig, H. Poizner, T. Sejnowski, G. Cauwenberghs, Closed-loop brain–machine-body interfaces for noninvasive rehabilitation of movement disorders, Annals of Biomedical Engineering 42 (8) (2014) 1573–1593.

[111] T. Denison, M. Morris, F. Sun, Building a bionic nervous system, IEEE Spectrum 52 (2) (2015) 32–39.

[112] P.R. Gigante, R.R. Goodman, Responsive neurostimulation for the treatment of epilepsy, Neurosurgery Clinics of North America 22 (4) (2011) 477–480.

[113] C.N. Heck, D. King-Stephens, A.D. Massey, D.R. Nair, B.C. Jobst, G.L. Barkley, V. Salanova, A.J. Cole, M.C. Smith, R.P. Gwinn, C. Skidmore, P.C. Van Ness, G.K. Bergey, Y.D. Park, I. Miller,

E. Geller, P.A. Rutecki, R. Zimmerman, D.C. Spencer, A. Goldman, J.C. Edwards, J.W. Leiphart, R.E. Wharen, J. Fessler, N.B. Fountain, G.A. Worrell, R.E. Gross, S. Eisenschenk, R.B. Duckrow, L.J. Hirsch, C. Bazil, C.A. O'Donovan, F.T. Sun, T.A. Courtney, C.G. Seale, M.J. Morrell, Two-year seizure reduction in adults with medically intractable partial onset epilepsy treated with responsive neurostimulation: final results of the RNS system pivotal trial, Epilepsia 55 (3) (2014) 432–441.

[114] M.J. Cook, T.J. O'Brien, S.F. Berkovic, M. Murphy, A. Morokoff, G. Fabinyi, W. D'Souza, R. Yerra, J. Archer, L. Litewka, S. Hosking, P. Lightfoot, V. Ruedebusch, W.D. Sheffield, D. Snyder, K. Leyde, D. Himes, Prediction of seizure likelihood with a long-term, implanted seizure advisory system in patients with drug-resistant epilepsy: a first-in-man study, Lancet Neurology 12 (6) (2013) 563–571.

[115] M. Bandarabadi, A. Dourado, A robust low complexity algorithm for real-time epileptic seizure detection, Epilepsia 55 (2014) 137.

[116] S. Ramgopal, S. Thome-Souza, M. Jackson, N.E. Kadish, I.S. Fernandez, J. Klehm, W. Bosl, C. Reinsberger, S. Schachter, T. Loddenkemper, Seizure detection, seizure prediction, and closed-loop warning systems in epilepsy, Epilepsy & Behavior 37 (2014) 291–307.

[117] Z. Zhang, K.K. Parhi, Low-complexity seizure prediction from iEEG/sEEG using spectral power and ratios of spectral power, IEEE Transactions on Biomedical Circuits and Systems 10 (3) (2016) 693–706.

[118] K. Abdelhalim, H.M. Jafari, L. Kokarovtseva, J.L. Perez Velazquez, R. Genov, 4-Channel UWB wireless neural vector analyzer SoC with a closed-loop phase synchrony-triggered neurostimulator, IEEE Journal of Solid-State Circuits 48 (10) (2013) 2494–2510.

[119] A. Page, C. Sagedy, E. Smith, N. Attaran, T. Oates, T. Mohsenin, A flexible multichannel EEG feature extractor and classifier for seizure detection, IEEE Transactions on Circuits and Systems II: Express Briefs 62 (2) (2015) 109–113.

[120] A.M. Bin Altaf, Z. Chen, J. Yoo, A 16-channel patient-specific seizure onset and termination detection SoC with impedance-adaptive transcranial electrical stimulator, IEEE Journal of Solid-State Circuits 50 (11) (2015) 2728–2740.

[121] G. Schalk, E.C. Leuthardt, Brain–computer interfaces using electrocorticographic signals, IEEE Reviews in Biomedical Engineering 4 (2011) 140–154.

[122] A.T. Avestruz, W. Santa, D. Carlson, R. Jensen, S. Stanslaski, A. Helfenstine, T. Denison, A 5 μW/Channel spectral analysis IC for chronic bidirectional brain–machine interfaces, IEEE Journal of Solid-State Circuits 43 (12) (2008) 3006–3024.

[123] F. Zhang, A. Mishra, A.G. Richardson, B. Otis, A low-power ECoG/EEG processing IC with integrated multiband energy extractor, IEEE Transactions on Circuits and Systems I: Regular Papers 58 (9) (2011) 2069–2082.

[124] C.S. Mestais, G. Charvet, F. Sauter-Starace, M. Foerster, D. Ratel, A.L. Benabid, WIMAGINE: wireless 64-channel ECoG recording implant for long term clinical applications, IEEE Transactions on Neural Systems and Rehabilitation Engineering 23 (1) (2015) 10–21.

[125] A.M. Sodagar, G.E. Perlin, Y. Ying, K. Najafi, K.D. Wise, An implantable 64-channel wireless microsystem for single-unit neural recording, IEEE Journal of Solid-State Circuits 44 (9) (2009) 2591–2604.

[126] A.M. Sodagar, K.D. Wise, K. Najafi, A wireless implantable microsystem for multichannel neural recording, IEEE Transactions on Microwave Theory and Techniques 57 (10) (2009) 2565–2573.

[127] F. Goodarzy, E. Skafidas, S. Gambini, Feasibility of energy-autonomous wireless micro-sensors for biomedical applications: powering and communication, IEEE Reviews in Biomedical Engineering 8 (2015) 17–29.

[128] A. Kim, M. Ochoa, R. Rahimi, B. Ziaie, New and emerging energy sources for implantable wireless microdevices, IEEE Access 3 (2015) 89–98.

[129] K. Murakawa, M. Kobayashi, O. Nakamura, S. Kawata, A wireless near-infrared energy system for medical implants, IEEE Engineering in Medicine and Biology Magazine 18 (6) (1999) 70–72.

[130] K. Goto, T. Nakagawa, O. Nakamura, S. Kawata, An implantable power supply with an optically rechargeable lithium battery, IEEE Transactions on Biomedical Engineering 48 (7) (2001) 830–833.

[131] C. Algora, R. Pěna, Recharging the battery of implantable biomedical devices by light, Artificial Organs 33 (10) (2009) 855–860.

[132] T.-C. Chou, R. Subramanian, J. Park, P.P. Mercier, A miniaturized ultrasonic power delivery system, in: Proceedings of the IEEE Biomedical Circuits and Systems Conference, 2014, pp. 440–443.

[133] D. Seo, T. Hao-Yen, J.M. Carmena, J.M. Rabaey, E. Alon, B.E. Boser, M.M. Maharbiz, Ultrasonic beamforming system for interrogating multiple implantable sensors, in: Proceedings of the Annual International Conference of the IEEE Engineering in Medicine and Biology Society, 2015, pp. 2673–2676.

[134] F.-G. Zeng, S. Rebscher, W. Harrison, X. Sun, H. Feng, Cochlear implants: system design, integration, and evaluation, IEEE Reviews in Biomedical Engineering 1 (2008) 115–142.

[135] H.-M. Lee, K.Y. Kwon, L. Wen, M. Ghovanloo, A power-efficient switched-capacitor stimulating system for electrical/optical deep brain stimulation, IEEE Journal of Solid-State Circuits 50 (1) (2015) 360–374.

[136] B.M. Badr, R. Somogyi-Gsizmazia, K.R. Delaney, N. Dechev, Wireless power transfer for telemetric devices with variable orientation, for small rodent behavior monitoring, IEEE Sensors Journal 15 (4) (2015) 2144–2156.

[137] R. Jegadeesan, S. Nag, K. Agarwal, N.V. Thakor, G. Yong-Xin, Enabling wireless powering and telemetry for peripheral nerve implants, IEEE Journal of Biomedical and Health Informatics 19 (3) (2015) 958–970.

[138] O. Knecht, R. Bosshard, J.W. Kolar, High-efficiency transcutaneous energy transfer for implantable mechanical heart support systems, IEEE Transactions on Power Electronics 30 (11) (2015) 6221–6236.

[139] M.W. Baker, R. Sarpeshkar, Feedback analysis and design of RF power links for low-power bionic systems, IEEE Transactions on Biomedical Circuits and Systems 1 (1) (2007) 28–38.

[140] U.-M. Jow, M. Ghovanloo, Design and optimization of printed spiral coils for efficient transcutaneous inductive power transmission, IEEE Transactions on Biomedical Circuits and Systems 1 (3) (2007) 193–202.

[141] A.K. RamRakhyani, S. Mirabbasi, C. Mu, Design and optimization of resonance-based efficient wireless power delivery systems for biomedical implants, IEEE Transactions on Biomedical Circuits and Systems 5 (1) (2011) 48–63.

[142] R.F. Xue, K.W. Cheng, M. Je, High-efficiency wireless power transfer for biomedical implants by optimal resonant load transformation, IEEE Transactions on Circuits and Systems I: Regular Papers 60 (4) (2013) 867–874.

[143] I. Mayordomo, T. Drager, P. Spies, J. Bernhard, A. Pflaum, An overview of technical challenges and advances of inductive wireless power transmission, Proceedings of the IEEE 101 (6) (2013) 1302–1311.

[144] H. Xu, J. Handwerker, M. Ortmanns, Telemetry for implantable medical devices: Part 2 – power telemetry, IEEE Solid-State Circuits Magazine 6 (3) (2014) 60–63.

[145] J.P. Reilly, Applied Bioelectricity, from Electrical Stimulation to Electropathology, Springer-Verlag, New York, 1998.

[146] D. Miklavčič, N. Pavšelj, F.X. Hart, Electric properties of tissues, in: Wiley Encyclopedia of Biomedical Engineering, John Wiley & Sons, Inc., 2006.

[147] C. Gabriel, S. Gabriel, Compilation of the dielectric properties of body tissues at RF and microwave frequencies, [Online]. Available: http://niremf.ifac.cnr.it/docs/DIELECTRIC/home.html.

[148] D. Ahn, M. Ghovanloo, Optimal design of wireless power transmission links for millimeter-sized biomedical implants, IEEE Transactions on Biomedical Circuits and Systems 10 (1) (2016) 125–137.

[149] M. Zargham, P.G. Gulak, Fully integrated on-chip coil in 0.13 μm CMOS for wireless power transfer through biological media, IEEE Transactions on Biomedical Circuits and Systems 9 (2) (2015) 259–271.

[150] A.S.Y. Poon, S. O'Driscoll, T.H. Meng, Optimal frequency for wireless power transmission into dispersive tissue, IEEE Transactions on Antennas and Propagation 58 (5) (2010) 1739–1750.

[151] M. Zargham, P.G. Gulak, Maximum achievable efficiency in near-field coupled power-transfer systems, IEEE Transactions on Biomedical Circuits and Systems 6 (3) (2012) 228–245.

[152] P.P. Mercier, A.P. Chandrakasan (Eds.), Ultra-Low-Power Short-Range Radios, Springer, 2015.

[153] A.P. Chandrakasan, F.S. Lee, D.D. Wentzloff, V. Sze, B.P. Ginsburg, P.P. Mercier, D.C. Daly, R. Blazquez, Low-power impulse UWB architectures and circuits, Proceedings of the IEEE 97 (2) (2009) 332–352.

[154] P.P. Mercier, D.C. Daly, A.P. Chandrakasan, An energy-efficient all-digital UWB transmitter employing dual capacitively-coupled pulse-shaping drivers, IEEE Journal of Solid-State Circuits 44 (6) (2009) 1679–1688.

[155] H. Ando, K. Takizawa, T. Yoshida, K. Matsushita, M. Hirata, T. Suzuki, Wireless multichannel neural recording with a 128 Mbps UWB transmitter for implantable brain–machine interfaces, IEEE Transactions on Biomedical Circuits and Systems 10 (6) (2016) 1068–1078.

[156] S.A. Mirbozorgi, H. Bahrami, M. Sawan, L.A. Rusch, B. Gosselin, A single-chip full-duplex high speed transceiver for multi-site stimulating and recording neural implants, IEEE Transactions on Biomedical Circuits and Systems 10 (3) (2016) 643–653.

[157] Z. Tang, B. Smith, J.H. Schild, P.H. Peckham, Data transmission from an implantable biotelemeter by load-shift keying using circuit configuration modulator, IEEE Transactions on Biomedical Engineering 42 (5) (1995) 524–528.

[158] C. Sauer, M. Stanacevic, G. Cauwenberghs, N. Thakor, Power harvesting and telemetry in CMOS for implanted devices, IEEE Transactions on Circuits and Systems I: Regular Papers 52 (12) (2005) 2605–2613.

[159] S. Mandal, R. Sarpeshkar, Power-efficient impedance-modulation wireless data links for biomedical implants, IEEE Transactions on Biomedical Circuits and Systems 2 (4) (2008) 301–315.

[160] H.-M. Lee, M. Ghovanloo, An integrated power-efficient active rectifier with offset-controlled high speed comparators for inductively powered applications, IEEE Transactions on Circuits and Systems I: Regular Papers 58 (8) (2011) 1749–1760.

[161] S. Ha, C. Kim, J. Park, S. Joshi, G. Cauwenberghs, Energy-recycling integrated 6.78-Mbps data 6.3-mW power telemetry over a single 13.56-MHz inductive link, in: Symposium on VLSI Circuits Digest of Technical Papers, IEEE, 2014, pp. 66–67.

[162] B.S. Wilson, M.F. Dorman, Cochlear implants: a remarkable past and a brilliant future, Hearing Research 242 (1–2) (2008) 3–21.

[163] H. Bhamra, Y. Kim, J. Joseph, J. Lynch, O.Z. Gall, H. Mei, C. Meng, J. Tsai, P. Irazoqui, A 24 μW batteryless, crystal-free, multinode synchronized SoC "Bionode" for wireless prosthesis control, IEEE Journal of Solid-State Circuits 50 (11) (2015) 2714–2727.

[164] A. Yakovlev, J.H. Jang, D. Pivonka, An 11 μW sub-pJ/bit reconfigurable transceiver for mm-sized wireless implants, IEEE Transactions on Biomedical Circuits and Systems 10 (1) (2016) 175–185.

[165] Y.P. Lin, C.Y. Yeh, P.Y. Huang, Z.Y. Wang, H.H. Cheng, Y.T. Li, C.F. Chuang, P.C. Huang, K.T. Tang, H.P. Ma, Y.C. Chang, S.R. Yeh, H. Chen, A battery-less, implantable neuro-electronic interface for studying the mechanisms of deep brain stimulation in rat models, IEEE Transactions on Biomedical Circuits and Systems 10 (1) (2016) 98–112.

[166] Z. Xiao, X. Tan, X. Chen, S. Chen, Z. Zhang, H. Zhang, J. Wang, Y. Huang, P. Zhang, L. Zheng, H. Min, An implantable RFID sensor tag toward continuous glucose monitoring, IEEE Journal of Biomedical and Health Informatics 19 (3) (2015) 910–919.

[167] H. Yu, K. Najafi, Low-power interface circuits for bio-implantable microsystems, in: IEEE International Solid-State Circuits Conference Digest of Technical Papers, 2003, pp. 194–487.

[168] H.-M. Lee, H. Park, M. Ghovanloo, A power-efficient wireless system with adaptive supply control for deep brain stimulation, IEEE Journal of Solid-State Circuits 48 (9) (2013) 2203–2216.

[169] G. Jiang, D.D. Zhou, Technology advances and challenges in hermetic packaging for implantable medical devices, in: D. Zhou, E. Greenbaum (Eds.), Implantable Neural Prostheses 2: Techniques and Engineering Approaches, Springer New York, New York, NY, 2010, pp. 27–61.

[170] A. Vanhoestenberghe, N. Donaldson, Corrosion of silicon integrated circuits and lifetime predictions in implantable electronic devices, Journal of Neural Engineering 10 (3) (2013) 031002.

[171] F.T. Sun, M.J. Morrell, R.E. Wharen, Responsive cortical stimulation for the treatment of epilepsy, Neurotherapeutics 5 (1) (2008) 68–74.

[172] NeuroPace RNS System, http://www.neuropace.com.

[173] D. Prutchi, NeuroVista Publishes Study Results for their Implantable Seizure-Warning Device, http://www.implantable-device.com.

[174] K.A. Davis, B.K. Sturges, C.H. Vite, V. Ruedebusch, G. Worrell, A.B. Gardner, K. Leyde, W.D. Sheffield, B. Litt, A novel implanted device to wirelessly record and analyze continuous intracranial canine EEG, Epilepsy Research 96 (1–2) (2011) 116–122.

[175] A.G. Rouse, S.R. Stanslaski, P. Cong, R.M. Jensen, P. Afshar, D. Ullestad, R. Gupta, G.F. Molnar, D.W. Moran, T.J. Denison, A chronic generalized bi-directional brain–machine interface, Epilepsy Research 8 (3) (2011) 036018.

[176] P. Afshar, A. Khambhati, S. Stanslaski, D. Carlson, R. Jensen, D. Linde, S. Dani, M. Lazarewicz, P. Cong, J. Giftakis, P. Stypulkowski, T. Denison, A translational platform for prototyping closed-loop neuromodulation systems, Frontiers in Neural Circuits 6 (2013).

[177] M. Hirata, T. Yoshimine, Electrocorticographic brain–machine interfaces for motor and communication control, in: Clinical Systems Neuroscience, 2015.

[178] A.C. Ho, M.S. Humayun, J.D. Dorn, L. da Cruz, G. Dagnelie, J. Handa, P.O. Barale, J.A. Sahel, P.E. Stanga, F. Hafezi, A.B. Safran, J. Salzmann, A. Santos, D. Birch, R. Spencer, A.V. Cideciyan, E. de Juan, J.L. Duncan, D. Eliott, A. Fawzi, L.C. Olmos de Koo, G.C. Brown, J.A. Haller, C.D. Regillo, L.V. Del Priore, A. Arditi, D.R. Geruschat, R.J. Greenberg, Long-term results from an epiretinal prosthesis to restore sight to the blind, Ophthalmology 122 (8) (2015) 1547–1554.

[179] A.A. Weaver, K.L. Loftis, J.C. Tan, S.M. Duma, J.D. Stitzel, CT based three-dimensional measurement of orbit and eye anthropometry, Investigative Ophthalmology and Visual Science 51 (10) (2010) 4892–4897.

[180] J. Jeong, S.H. Bae, K.S. Min, J.-M. Seo, H. Chung, S.J. Kim, A miniaturized, eye-conformable, and long-term reliable retinal prosthesis using monolithic fabrication of liquid crystal polymer (LCP), IEEE Transactions on Biomedical Engineering 62 (3) (2015) 982–989.

[181] J.D. Weiland, M.S. Humayun, Retinal prosthesis, IEEE Transactions on Biomedical Engineering 61 (5) (2014) 1412–1424.

[182] J.D. Weiland, F.M. Kimock, J.E. Yehoda, E. Gill, B.P. McIntosh, P.J. Nasiatka, A.R. Tanguay, Chip-scale packaging for bioelectronic implants, in: Proceedings of the International IEEE/EMBS Conference on Neural Engineering, 2013, pp. 931–936.

[183] S.-W. Hwang, H. Tao, D.-H. Kim, H. Cheng, J.-K. Song, E. Rill, M.A. Brenckle, B. Panilaitis, S.M. Won, Y.S. Kim, Y.M. Song, K.J. Yu, A. Ameen, R. Li, Y. Su, M. Yang, D.L. Kaplan, M.R. Zakin, M.J. Slepian, Y. Huang, F.G. Omenetto, J.A. Rogers, A physically transient form of silicon electronics, Science 337 (6102) (2012) 1640–1644.

[184] K.J. Yu, D. Kuzum, S.-W. Hwang, B.H. Kim, H. Juul, N.H. Kim, S.M. Won, K. Chiang, M. Trumpis, A.G. Richardson, H. Cheng, H. Fang, M. Thompson, H. Bink, D. Talos, K.J. Seo, H.N. Lee, S.-K. Kang, J.-H. Kim, J.Y. Lee, Y. Huang, F.E. Jensen, M.A. Dichter, T.H. Lucas, J. Viventi, B. Litt, J.A. Rogers, Bioresorbable silicon electronics for transient spatiotemporal mapping of electrical activity from the cerebral cortex, Nature Materials 15 (2016) 782–791.

CHAPTER 2

Integrated circuit interfaces for ECoG signal recording

Contents

2.1. Introduction

Physiological signal monitoring technology has advanced tremendously over the years, making a substantial impact on medical diagnostics and personal healthcare, from the early fundamental advances in biopotential sensing technology [1–3] to most recent advances extending the range of physiological sensing using sensing technology abundantly available in handheld devices and household appliances [4–7]. CMOS technologies and circuit techniques have facilitated the development and miniaturization of innovative physiological sensing devices, improving the performance, power and monetary costs while ensuring the validity of medical information through analog and digital signal processing methods. These IC developments have permitted reliable measurement of brain signals and have spawned a variety of new instruments for clinical treatment and diagnosis.

Innovations by semiconductor technologies enable ambulatory continuous-time monitoring of patients even at home. This ubiquitous monitoring supported by modern IC technology can enable personalized healthcare and preemptive medicine, which are emerging solutions to soaring healthcare costs induced by the current demographical trend of increasing aging population. The patient-supporting sensors and systems not

High-Density Integrated Electrocortical Neural Interfaces
https://doi.org/10.1016/B978-0-12-815115-0.00009-X

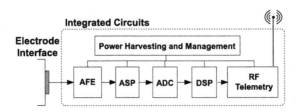

Figure 2.1 A generic block diagram of an integrated circuit (IC) and electrode interface for ECoG signal acquisition, coding, and transmission.

only extend the capability and accuracy of modern diagnostics, but also improve the patient's everyday life.

Fig. 2.1 shows the main functional components of a generic IC for ECoG signal monitoring, comprising analog frontend (AFE), analog signal processor (ASP), analog-to-digital converter (ADC), digital signal processor (DSP), radio frequency (RF) communications, and power management. This chapter surveys these components with a focus on the core functions of AFE, ASP, and ADC implemented in low-noise, low-power custom integrated circuits, and tailored to the signal conditions and range of the physiological variables of interest. Foremost, a solid and thorough understanding of the electrode–body interface is of primary importance for accurate and reliable ECoG sensing and signal acquisition.

2.2. Requirements for ECoG signal recording

2.2.1 Physiological requirements

The main design requirements of the AFE, ASP, and ADC are driven by characteristics of the physiological signals and the body–electrode interface. Biopotentials, such as EEG, ECG, EMG and ECoG, are generated from volume conduction of currents made by collections of electrogenic cells. EEG is the electrical potential induced from collective activities of large number of neurons in the brain. ECG results from action potentials of cardiac muscle cells, and EMG from contractions of skeletal muscle cells. Various other biopotentials (EOG, ERG, EGG, etc.) also result from collective effects of large numbers of electrogenic cells or ionic distribution.

As shown in Fig. 2.2, ECoG signals range over very low frequency, typically less than 1 kHz. They are very low in amplitude ranging from tens to hundreds μV when measured by a surface electrode. Since ECoG range down to less than 1 Hz, recording of these signals faces challenges in electrode offset voltage, which may reach up to 100 mV, varying slowly over time. In addition, $1/f$ noise needs to be suppressed if the application calls for low noise at low frequencies (< 1 Hz). Also, common-mode interference from the mains and other irrelevant biopotentials should be sufficiently rejected.

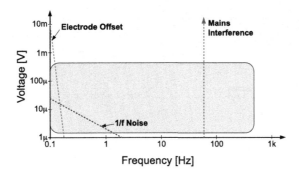

Figure 2.2 Characteristics of ECoG signal, in relation to mains interference, electrode offset drift and $1/f$ noise [8,9].

2.2.2 Design factors, inter-relations, and trade-offs

Several factors quantifying metrics of performance and cost in the design, their inter-relationships, and typical ranges from the literature, are summarized in Table 1.1. Several of these relationships, such as between noise, power, bandwidth, gain, and dynamic range are generally well understood deriving from fundamental physical and information-theoretic principles, e.g., power is typically linear in bandwidth but subject to noise considerations. Other relationships, such as between input impedance and movement artifact rejection, are specific to the physiological signals and environmental factors at the electrode interface. The various intertwined relationships between these factors must be co-optimized in the design trade-offs at the electrode, circuit, and architectural levels. A deep understanding of fundamental principles linking these factors and driving the trade-offs is thus required. Specific trade-offs and architectural design topologies that take advantage of properties of low-power CMOS integrated circuits and systems are elaborated in the following sections.

2.3. Subthreshold operation of MOS transistors

Fig. 2.3(A) illustrates a cross-section of fabricated NMOS and PMOS field-effect transistors in an integrated circuit CMOS process. Both have four terminals: gate (G), source (S), drain (D), and body (B). Their schematic drawing symbols are depicted in Fig. 2.3(B). MOSFET operation can be separated into two modes according to the voltages between gate and source terminals: strong inversion and weak inversion (or subthreshold) operation as shown in Fig. 2.3(C).

Counter to standard practices in analog CMOS circuit design, the weak inversion (subthreshold) region of CMOS operation has proven a favorable regime for low-power biomedical circuit design. In conventional design and particularly for high-speed applications, weak inversion operation has been considered as non-ideality in the cut-off

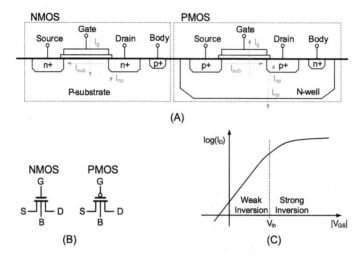

Figure 2.3 (A) Cross-section of NMOS and PMOS FETs fabricated in a CMOS process, (B) their schematic drawing symbols and (C) logarithm of drain current ID as a function of gate-source voltage with body tied to source, and drain voltage biased in the saturation region.

region and its current has been labeled as leakage current. Recently, weak inversion has become increasingly important because its low power and low bandwidth characteristics are well suited for biomedical and other low-power sensor applications, owing to superior transconductance efficiency. Furthermore, transistors in deep sub-micron technology operating in weak inversion do not suffer from many process-dependent problems plaguing the above-threshold strong inversion region, such as gain-limiting effects of velocity saturation in electron and hole mobility [10].

Transistor model equations in weak inversion are simpler, are more transparent, and scale over a wider range than in strong inversion. The electron energy of a transistor in weak inversion is based entirely on the Boltzmann distribution, independent of process technology. The drain current through the transistor channel flows not by drift, but by diffusion, and changes exponentially with gate voltage. Thus, weak inversion operation is particularly well suited for implementing translinear circuits and log-domain filters.

Drain current I_{DS}, transconductance g_m and unity-gain frequency f_t in weak inversion are as follows [11]:

$$i_{DS} = i_{DS0} \frac{W}{L} e^{v_{GS}/(nV_t)} \left(1 - e^{-v_{DS}/V_t}\right), \tag{2.1}$$

$$g_m = \frac{I_{DS}}{nV_t}, \tag{2.2}$$

$$f_t = \frac{I_{DS}}{2\pi n V_t \left(C_{gs} + C_{gd} + C_{gb}\right)} \propto I_{DS}. \tag{2.3}$$

Because the transconductance is linearly proportional to drain current, so is unity-gain frequency. Thus, the trade-off between current and bandwidth is very straightforward: the larger the current, the wider the bandwidth.

Thermal noise in saturation and weak inversion is proportional to drain current as follows [12]:

$$\overline{i_{n,th}^2} = 2q\overline{I_{DS}}\Delta f \tag{2.4}$$

where Δf is the signal bandwidth. The relative noise power (inverse of the signal-to-noise ratio) is inversely proportional to drain current:

$$\frac{\overline{i_{n,th}^2}}{I_{DS}^2} = \frac{2q\Delta f}{\overline{I_{DS}}}. \tag{2.5}$$

Therefore, signal-to-noise ratio is linearly proportional to bias current in weak inversion. For a majority of biomedical applications with narrow signal bandwidth, the lower currents of circuits in weak inversion still offer adequately large signal-to-noise ratio at maximum energy efficiency.

Flicker noise, also known as $1/f$ noise or pink noise, is also a significant noise source at low frequency. Random captures of carriers in traps near the Si/SiO_2 interface and some other mechanisms are known to be a main source of $1/f$ noise [13,14], which is given by

$$\overline{i_{n,f}^2} = \frac{g_m^2 K}{C_{ox} WL} \cdot \frac{1}{f}\Delta f \tag{2.6}$$

where K is a process-dependent constant, W and L are width and length of the MOS transistor, and C_{ox} is the gate oxide capacitance. PMOS transistors are known to have less $1/f$ noise than NMOS transistors, and therefore should be used in the input differential pair of a frontend amplifier for low-noise low-frequency applications in biosensing. Enlarging the MOS device size also decreases $1/f$ noise inversely proportional to area.

The $1/f$ noise corner frequency f_c serves as an important indication for the proportion of $1/f$ noise relative to thermal noise over the spectrum. The $1/f$ noise corner f_c is defined as the frequency at which $1/f$ noise and thermal noise are at equal magnitude, which in weak inversion is given by:

$$\overline{i_{n,th}^2} = \overline{i_{n,f}^2}, \tag{2.7}$$

$$2nkTg_m\Delta f = \frac{g_m^2 K}{C_{ox} WL} \cdot \frac{1}{f}\Delta f, \tag{2.8}$$

$$f_c = \frac{K}{C_{ox} WL}g_m\frac{1}{2nkT}. \tag{2.9}$$

The $1/f$ noise corner can vary from a few 100 Hz to a few MHz depending on quality of process fabrication. Also, it depends on bias current: the lower the bias current, the lower the $1/f$ noise corner and hence the smaller the relative $1/f$ contribution to the overall noise.

For low-noise biomedical applications such as EEG acquisition, chopper stabilization techniques are widely used to reduce $1/f$ noise further. To be effective, the chopping frequency needs to be significantly higher than the $1/f$ noise corner frequency, typically ranging between 100 Hz and 10 kHz in these applications. Other techniques such as auto-zeroing and correlated double sampling can be used to reduce $1/f$ noise as well.

Three types of currents in MOS transistors need careful consideration in biomedical circuit design, below. These three are depicted in Fig. 2.3(A) as subthreshold current I_{sub}, gate leakage I_g, and p–n-junction reverse-bias leakage I_{np} [15].

2.4. Basic architectures of instrumentation amplifiers

One of the most challenging parts in the design of wearable physiological monitoring system is the implementation of instrumentation amplifiers (IAs), which acquire biosignals from electrodes and perform analog signal processing and conditioning. IAs are subject to almost all the challenging design specifications aforementioned in Sect. 2.2.2.

A classic 3-Op-Amp IA is adequate for achieving large input impedance, large CMRR and sufficient gain. However, it consumes large power and area [16] since it uses three amplifiers. For a micropower biopotential acquisition frontend, the configurations shown in Fig. 2.4(A) (similar to [17]) and (B) (similar to [18]) are widely used. The AC-coupling input capacitors C_C block electrode offset voltages. Owing to favorable matching performance of capacitors in integrated CMOS processes, the gain can be precisely controlled. A large resistor R_f, typically implemented by a pseudoresistor or a switched-capacitor circuit (Sect. 2.6.2), establishes DC biasing of the voltage at the input nodes of the amplifier and performs highpass filtering together with C_f. Mismatch in capacitor values results in degradation of CMRR. A practical CMRR that this architecture can achieve is about 60 to 70 dB. In addition, C_C dominates the input impedance. Therefore, the value of C_C needs to be set by considering CMRR and the input impedance.

The transfer function of the IA is as follows [17]:

$$A_v(s) = -\frac{C_C}{C_f} \cdot \frac{1 - \frac{sC_f}{g_m}}{\left(1 + \frac{1}{sR_f C_f}\right) \cdot \left(1 + \frac{s(C_C + C_L + C_C C_L / C_f)}{g_m}\right)} \tag{2.10}$$

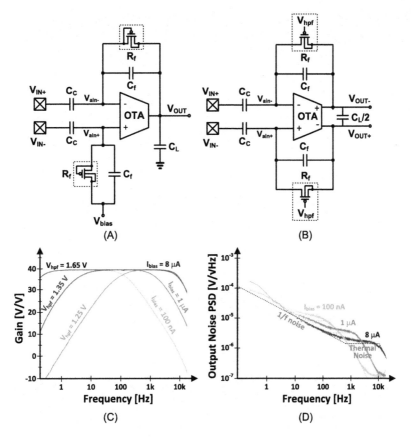

Figure 2.4 Generic architectures of (A) single-ended output [17], and (B) fully differential instrumentation amplifiers. Measured (C) transfer function and (D) output noise power spectral density for different configuration settings of the OTA bias current I_{bias} and pseudoresistance voltage bias V_{hpf} [18].

where g_m is transconductance of the OTA. Its passband gain is determined by the ratio of capacitors C_C to C_f. The highpass cutoff frequency f_{HP} is given by

$$f_{HP} = \frac{1}{2\pi R_f C_f}. \tag{2.11}$$

The lowpass cutoff frequency f_{LP} can be controlled by the load capacitor C_L and is approximately given by

$$f_{LP} \cong \frac{g_m C_f}{2\pi C_C C_L}, \tag{2.12}$$

when $C_L \gg C_f$ and $C_C/C_f \gg 1$. The right-half-plane zero at g_m/C_f can be canceled by inserting a $1/g_m$ resistor in series with C_f. However, it can be ignored in many cases because it is located at much higher frequency than the frequency range of interest for biomedical applications.

The main noise contributors of the single-stage IAs are the operational transconductance amplifier (OTA) and the feedback resistor R_f. The input-referred noises due to these sources are respectively given below:

$$\sqrt{\overline{v_{ni,R_f}^2}} = \sqrt{\frac{4kT}{R_f} \cdot \frac{1}{2\pi f C_{in}}}, \tag{2.13}$$

$$\sqrt{\overline{v_{ni,amp}^2}} = \sqrt{\overline{v_{ni,system}^2}} \cdot \left(\frac{C_f + C_{in} + C_C}{C_C} \right) \tag{2.14}$$

where C_{in} refers to parasitic capacitance at the OTA input nodes V_{IN+} and V_{IN-} [16]. The noise from R_f can be dominant at low frequencies of interest. Thus, the following criterion needs to be met in order to reduce the contribution of the noise from R_f [19]:

$$\frac{C_L}{C_C} \ll \frac{2}{3} \frac{f_{LP}}{f_{HP}}. \tag{2.15}$$

In practical circuits, the noise from the OTA is dominant over the noise from R_f [19].

As a benchmark in the design of frontend IAs for low noise and low supply current, the noise efficiency factor (NEF) is used to compare the current–noise performance:

$$\text{NEF} = V_{rms,in} = \sqrt{\frac{2I_{tot}}{\pi V_t \cdot 4kT \cdot BW}} \tag{2.16}$$

where V_{rms} is the total input-referred noise, I_{tot} the total current drain in the system, V_t the thermal voltage, and BW the -3-dB bandwidth of the system [20]. NEF corresponds to the normalized supply current relative to that of a single BJT with ideal current load for the same noise level, defining the theoretical limit (NEF = 1). In practice, differential IAs with input differential pairs incur twice the supply current for the same transconductance, with NEF values greater than 2. The state-of-the-art IAs typically have NEF of 2.5 to 10. Recently, IAs using current-reuse techniques to boost g_m of the input stage have been demonstrated with NEFs below 2 [21–24]. As a point of reference, measured NEF and performance of state-of-the-art IAs are shown in Figs. 1.6 [25,17,18,26–37].

Note that even though NEF is widely used to benchmark IAs, NEF is nothing more but a trade-off between only three performance metrics: bandwidth, noise and current—excluding many other critical performance factors such as CMRR, input impedance, power consumption, input dynamic range, etc. Modified NEF metrics, one comparing power consumption instead of current [38], and another including power consumption and dynamic range [39], have been proposed as more comprehensive NEF alternatives.

2.5. Basic amplifier design techniques

The operational transconductance amplifier (OTA) is the most important block in an IA, as shown centrally in Fig. 2.4(A) (single-ended) and (B) (fully differential). Among many kinds of amplifier, the most popular two amplifiers, symmetrical OTA (or current mirror OTA) and folded cascode OTA, are shown in Fig. 2.5 [17,33]. Both are one-stage amplifiers since both have only one high-impedance node at V_{OUT}. The gain of both amplifiers is the product of the transconductance of the input pair and the output impedance. Fully differential amplifiers (not shown here) typically add more complexity in the design with the need for common-mode feedback, but offer superior CMRR and PSRR over single-ended solutions.

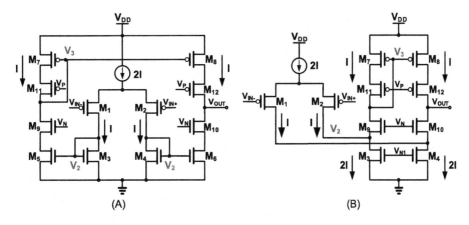

Figure 2.5 (A) Symmetrical OTA. (B) Folded cascode OTA.

2.5.1 Noise–power trade-off

Noise of the frontend amplifier for biomedical applications is critical. With given transconductances gm1 for M_1–M_2, g_{m3} for M_3–M_6 and g_{m7} for M_7–M_8, the input-referred thermal noise for the each amplifier in Fig. 2.5 is given respectively as follows [17]:

$$\overline{v_{ni,sym}^2} = \frac{16kT}{3g_{m1}}\left(\frac{3}{4}n + 2\frac{g_{m3}}{g_{m1}} + \frac{g_{m7}}{g_{m1}}\right)\Delta f, \tag{2.17}$$

$$\overline{v_{ni,folded}^2} = \frac{16kT}{3g_{m1}}\left(\frac{3}{4}n + \frac{g_{m3}}{g_{m1}} + \frac{g_{m7}}{g_{m1}}\right)\Delta f \tag{2.18}$$

where n is dependent on the capacitance ratio between depletion and oxide capacitance, which is generally between 1.2 and 1.8. In these equations, M_1 and M_2 are assumed in the weak inversion region while the others are in the strong inversion region. The effect from cascode MOSFETs are negligible in terms of noise. These equations clearly show that in order to obtain good noise performance, g_{m1} should be as large as possible. With a given bias current, the input pair transistors (M_1 and M_2) should operate in the weak inversion for higher g_m/I_D efficiency. In contrast, M_3–M_8 need to be in the strong inversion region in order to get lower g_m. Due to the approximation that g_{m1} is much larger than g_{m3} and g_{m7}, the input-referred noise is simplified as

$$\overline{v_{ni,sym}^2} \cong \overline{v_{ni,folded}^2} \cong \frac{4nkT}{g_{m1}}\Delta f = \frac{4n^2kTV_t}{I_1}\Delta f \propto \frac{1}{\text{current}}. \tag{2.19}$$

With a proper sizing in the amplifier, there is only one factor that can be manipulated by circuit designers – the current. Therefore, the trade-off between power consumption and noise is straightforward: the larger the current, the lower the noise. For more stringent constraint in power consumption, the telescopic amplifier is superior to the both amplifiers because it has only two branches. However, output swing range is much narrower than the two amplifiers aforementioned.

2.5.2 The g_m/I_D design methodology

The g_m/I_D design methodology is very practical for low-power and low-noise design [40]. The transconductance efficiency g_m/I_D indicates the efficiency of a transistor in achieving higher gain with lower current; g_m/I_D is larger in the weak inversion than in the strong inversion as shown in Fig. 2.6(A). It means that a transistor operates more efficiently in weak inversion. As Fig. 2.6(B) depicts, the operation region of a transistor with a given bias current can be manipulated by sizing for desired transconductance efficiency. However, it should be noted that larger device size for higher transconductance efficiency consumes larger area inducing larger parasitic capacitances.

2.5.3 Stability

Negative feedback is utilized in almost all amplifiers to acquire a precise gain, set independent of the open-loop gain by a feedback ratio of linear passive components, along with widened bandwidth and increased noise suppression. However, feedback in high-gain systems may be subject to possible sources of instability, which requires careful design consideration. IAs for biomedical applications do not require wide bandwidth since the frequency range of interest is rather low, typically less than 1 kHz. However, due to low-power constraints, typical currents are a few tens to hundreds of nA, and transistors are typically sized large for better matching and low noise performance, resulting in low conductances over large parasitic capacitances, which may worsen the stability.

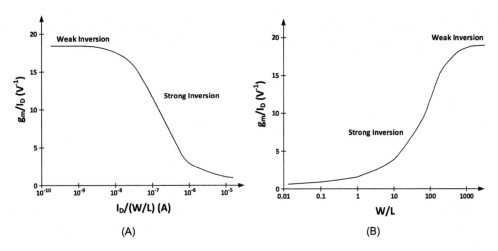

Figure 2.6 Transconductance efficiency (A) as a function of bias current with a fixed size, and (B) as a function of size with a given bias current.

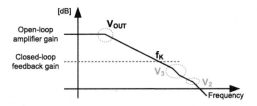

Figure 2.7 Open-loop Bode plot for the one-stage amplifiers.

Fig. 2.7 shows an open-loop Bode plot for the amplifiers. Since V_{OUT} is the only high–impedance node as mentioned above, the pole at V_{OUT} is located at very low frequency. The feedback establishes the closed loop gain indicated by the dashed horizontal line in Fig. 2.7. The intersection of the closed loop gain of the feedback and open loop gain of the amplifier determines the unity loop-gain frequency f_K. The phase margin is given by:

$$\text{Phase Margin} = 90 - \tan^{-1}\frac{f_K}{f_{p2}} \tag{2.20}$$

where f_{p2} is the second pole frequency. The position of the second pole is hence critical in the phase margin. If M_3–M_8 in Fig. 2.5(A)–(B) are assumed to have identical overdrive voltage, V_3 node has larger gate capacitance due to the larger device size, and may generate the second pole. However, this pole is always followed by a zero due to double signal paths, so it does not affect the phase margin seriously. Furthermore, in a fully differential amplifier, the pole–zero pair at the V_3 node is not induced. The pole at V_2 may be lower than the pole–zero pair due to the large junction capacitances of M_1 and

M_2, which are sized for large input-pair transconductance and low $1/f$ noise. In some cases, this may generate a doublet, causing relatively long settling time, although within acceptable range for many biomedical applications including ECoG recording.

2.6. Advanced techniques for instrumentation amplifier design

2.6.1 Offset and $1/f$ noise cancellation techniques

Autozeroing switched-capacitor techniques are often used to suppress electrode voltage offset and $1/f$ noise of the amplifier [40]. However, opening of the reset switch on the sampling capacitor after autozeroing introduces significant Nyquist–Johnson noise (kT/C noise) [41,42] and random charge injection that contaminate the sampled signal. The kT/C noise of a 1–10 pF capacitance alone is about tens of μV. To resolve this noise issue referred to the input of the AFE, signal folding and digital-assisted signal stitching can be used, resulting in relieving the specification of voltage dynamic range [43]. Instead, almost all of the IAs utilize chopper stabilization techniques to obviate $1/f$ noise.

The chopper modulation technique is widespread and essential to mitigate $1/f$ noise and other low-frequency noise, such as popcorn noise, voltage offsets and drifts. It is particularly used for sensitive acquisition of relatively weak biopotentials such as ECoG and EEG, which requires very low input-referred noise. The principles of the chopper modulation technique for amplifiers, which have been extensively studied [40,44–46], are illustrated in Fig. 2.8. The low-frequency band-limited input signal V_{in} is modulated in front of the amplifier by a square-wave chopping signal. The resulting waveform V_a for the signal is shifted to the chopping frequency f_{ch}, and the aggressors do not fall within the signal band. After the amplification and demodulation with the same chopping signal, the amplified input signal components are shifted to DC baseband frequency at V_b while the aggressors are moved to f_{ch} outside of the signal band. All the undesired aggressors and the harmonics are filtered out through the low-pass filter, and the desired input signal is ideally restored at the output V_{out}.

The residual offset is mainly caused by the non-idealities of the input chopper modulator. The mismatch of the clock-feedthrough and the charge injection in the input chopper generates switching transient spikes, which are demodulated at the output chopper into a residual output offset. In order to minimize the offset, at first, careful design and layout need to be done. A continuous [25,29,31,33] or digital [26] DC servo-loop can reduce the residual offset, and mitigate the signal distortion problem that is caused by the finite bandwidth of the amplifier. Alternatively, filtering techniques [30, 47,48] can be applied.

Output ripple is induced by the input offset of the amplifier, and can saturate the output of the amplifier since offset is also amplified. The ripple can be reduced by a continuous ripple-reduction loop [31] and a digital foreground calibration [26].

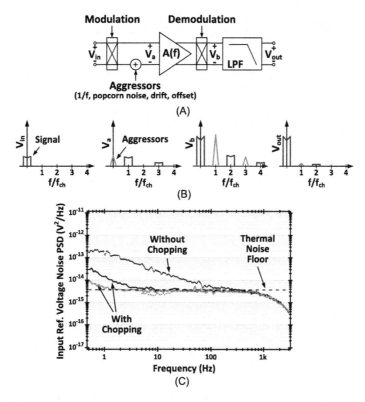

Figure 2.8 (A) Block diagram and (B) frequency-domain illustration of the chopper technique for low-frequency noise and drift cancellation [40,44–46]. (C) Input-referred noise spectrum with and without chopping [30].

2.6.2 Pseudoresistors for sub-Hz highpass cutoff

Highpass cutoff frequency needs to be well below 1 Hz in typical biomedical sensors, requiring ultra-high resistance in the TΩ range. Realizing sub-Hz time constants with on-chip capacitors and poly resistors consumes an impractically large area for integrated implementation.

The most prevalent solutions are combinations of 1–10 pF capacitors with PMOS-based MOS-bipolar pseudoresistors as shown in Fig. 2.9(A)–(F) [49,17]. The most basic topology among these is a PMOS whose gate and body terminals are connected as Fig. 2.9(A) [49]. This PMOS pseudoresistor combines a p–n-junction in the forward direction ($V_A > V_B$) with a diode-connected subthreshold PMOS in the reverse direction ($V_A < V_B$). Owing to the source-bulk connection, the gate-connected drain terminal is leakage free and is ideally connected to a leakage-sensitive side such as a floating input to an OTA. The measured resistance of a single MOS-bipolar pseudoresistor is shown in Fig. 2.9(I) [17]. A configuration with the PMOS gate connected to a bias voltage in

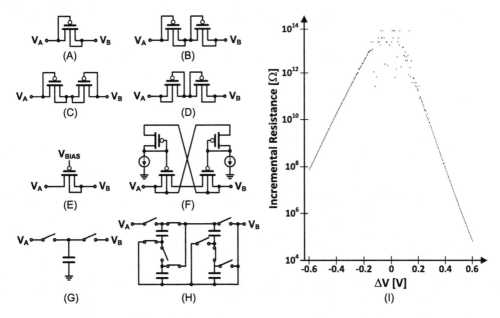

Figure 2.9 On-chip pseudoresistor implementations: (A) single MOS-bipolar pseudoresistor [49]; (B) pseudoresistor with two MOS-bipolar elements in series for twice higher resistance and greater voltage range [17,50]; (C) symmetrical version with outwardly connected gates; (D) symmetrical version with inwardly connected gates [28,26,51]; (E) voltage-biased pseudoresistor for resistance tunability [18,52–54,39]; (F) balanced tunable pseudoresistor with wider linear range [32]; (G) switched-capacitor implementation [29]; (H) switched-capacitor implementation with 10-times larger effective resistance than (g) [33]. (I) Measured resistance of a single MOS-bipolar pseudoresistor (A) as a function of voltage [17].

Fig. 2.9(E), results in a controllable resistance by gate voltage [18,52–54,39]. However, the resistance of the PMOS-based pseudoresistors in Fig. 2.9(A)–(E) drops drastically when the voltage across moves away from zero, inducing signal–dependent distortion while limiting the voltage dynamic range [32]. The pseudoresistor in Fig. 2.9(F) has balanced resistance with wider linear range up to a few hundred mV. An even wider linear range can be achieved by using an auxiliary amplifier [55].

However, standard MOS–bipolar pseudoresistors suffer from process, voltage and temperature (PVT) variations in addition to possible light and electromagnetic interference sensitivities, leading to variations in cutoff frequency. Switched capacitors can be used to implement on–chip PVT-insensitive high resistance as shown in Fig. 2.9(G) [29]. In this topology, switching frequency f_s and the capacitor in the middle determine the resistance precisely as $1/f_s C$. The switched-capacitor resistor in Fig. 2.9(H) mitigates manufacturability and interference issues by realizing a ten-fold resistance increase by charge sharing in the switched-capacitor circuits [33].

2.6.3 CMRR enhancement techniques

Common-mode interference is a difficult challenge for biomedical signal sensing systems. The major source of the interference comes from electric power lines, which are electrically coupled to the human body. High CMRR is required in the system to reject the common-mode interference in order to ensure high signal quality.

Accurate component matching between differential signal lines and between the channels is the most fundamental requirement to accomplish high CMRR. Good matching involves techniques from careful layout to smart architectural design choices.

2.6.3.1 Driven-right-leg technique

The DRL technique feeds the amplified input common-mode voltage into the body through an additional electrode, which has been placed on the right leg for ECG measurements. This negative feedback reduces the impedance in the feedback loop, attenuating the common-mode interference voltage at the sensor inputs [56,57] by factor of the feedback loop gain. Through the DRL negative feedback, the electrode impedance and common-mode voltage are reduced by the factor of the DRL loop gain. Hence in order to obtain higher gain, an open-loop DRL amplifier can be employed [58]. Digitally assisted DRL circuits offer larger gain at the mains frequency for higher rejection and lower gain elsewhere for stability [59]. In dry electrode applications, common-mode feedback to the negative inputs of individual amplifiers on the active electrodes also increases CMRR, and ensures stability unaffected by electrode impedance variations [26].

2.6.3.2 Input impedance boosting techniques

The variation and mismatch of electrode impedances also degrade CMRR, reduce signal amplitude, and make the system more susceptible to movement artifacts. Thus, the input impedance of the biopotential sensor should be much higher than the electrode impedance and the interface impedance between the body and the electrode. A positive feedback can bootstrap the AC-coupled input capacitors to boost the input impedance [60,26,34], achieving input impedance in the order of GΩ. In order to further boost input impedance to the TΩ level, a unity-gain amplifier with active shielding can be used to bootstrap capacitance of the input transistor and all other parasitic capacitance [61].

2.7. Conclusions

In this chapter we first presented system requirements for reliable ECoG signal acquisition. Then, we reviewed fundamental principles for low power integrated circuit design for ECoG signal sensing such as subthreshold and g_m/I_D design methods. Advanced techniques such as pseudoresistors, chopping, driven right leg circuits, and impedance bootstrapping were also covered.

References

[1] A. Lopez, P.C. Richardson, Capacitive electrocardiographic and bioelectric electrodes, IEEE Transactions on Biomedical Engineering BME-16 (1) (1969) 99.

[2] G.E. Bergey, R.D. Squires, W.C. Sipple, Electrocardiogram recording with pasteless electrodes, IEEE Transactions on Biomedical Engineering BME-18 (3) (1971) 206–2011.

[3] D.E. Hokanson, D.S. Sumner, D.E. Strandness, Electrically calibrated plethysmograph for direct measurement of limb blood-flow, IEEE Transactions on Biomedical Engineering Bm22 (1) (1975) 25–29.

[4] M. Garbey, N. Sun, A. Merla, I. Pavlidis, Contact-free measurement of cardiac pulse based on the analysis of thermal imagery, IEEE Transactions on Biomedical Engineering 54 (8) (2007) 1418–1426.

[5] M.Z. Poh, D.J. McDuff, R.W. Picard, Advancements in noncontact, multiparameter physiological measurements using a webcam, IEEE Transactions on Biomedical Engineering 58 (1) (2011) 7–11.

[6] C.G. Scully, J. Lee, J. Meyer, A.M. Gorbach, D. Granquist-Fraser, Y. Mendelson, K.H. Chon, Physiological parameter monitoring from optical recordings with a mobile phone, IEEE Transactions on Biomedical Engineering 59 (2) (2012) 303–306.

[7] O.T. Inan, D. Park, L. Giovangrandi, G.T.A. Kovacs, Noninvasive measurement of physiological signals on a modified home bathroom scale, IEEE Transactions on Biomedical Engineering 59 (8) (2012) 2137–2143.

[8] J.G. Webster, Medical Instrumentation: Application and Design, 4th ed., John Wiley & Sons, Inc., 2010.

[9] R.F. Yazicioglu, C. van Hoof, R. Puers, Biopotential Readout Circuits for Portable Acquisition Systems, Springer, 2009.

[10] C. Mead, Introduction to VLSI Systems, Addison Wesley, 1979.

[11] R. Sarpeshkar, Ultra Low Power Bioelectronics: Fundamentals, Biomedical Applications, and Bio-Inspired Systems, Cambridge University Press, 2010.

[12] R. Sarpeshkar, T. Delbruck, C.A. Mead, White noise in MOS transistors and resistors, IEEE Circuits and Devices Magazine 9 (6) (1993) 23–29.

[13] A.L. Mcwhorter, $1/f$ Noise and Related Surface Effects in Germanium, PhD dissertation, 1955.

[14] A. van der Ziel, Unified presentation of $1/f$ noise in electron devices: fundamental $1/f$ noise sources, Proceedings of the IEEE 76 (3) (1988) 233–258.

[15] K. Roy, S. Mukhopadhyay, H. Mahmoodi-Meimand, Leakage current mechanisms and leakage reduction techniques in deep-submicrometer CMOS circuits, Proceedings of the IEEE 91 (2) (2003) 305–327.

[16] M.J. Burke, D.T. Gleeson, A micropower dry-electrode ECG preamplifier, IEEE Transactions on Biomedical Engineering 47 (2) (2000) 155–162.

[17] R.R. Harrison, C. Charles, A low-power low-noise CMOS amplifier for neural recording applications, IEEE Journal of Solid-State Circuits 38 (6) (2003) 958–965.

[18] M. Mollazadeh, K. Murari, G. Cauwenberghs, N. Thakor, Micropower CMOS integrated low-noise amplification, filtering, and digitization of multimodal neuropotentials, IEEE Transactions on Biomedical Circuits and Systems 3 (1) (2009) 1–10.

[19] R.R. Harrison, The design of integrated circuits to observe brain activity, Proceedings of the IEEE 96 (7) (2008) 1203–1216.

[20] M.S.J. Steyaert, W.M.C. Sansen, Z.Y. Chang, A micropower low-noise monolithic instrumentation amplifier for medical purposes, IEEE Journal of Solid-State Circuits 22 (6) (1987) 1163–1168.

[21] B. Johnson, A. Molnar, An orthogonal current-reuse amplifier for multi-channel sensing, IEEE Journal of Solid-State Circuits 48 (6) (2013) 1487–1496.

[22] T. Yang, J. Holleman, An ultralow-power low-noise CMOS biopotential amplifier for neural recording, IEEE Transactions on Circuits and Systems II: Express Briefs 62 (10) (2015) 927–931.

[23] X. Liu, M. Zhang, T. Xiong, A.G. Richardson, T.H. Lucas, P.S. Chin, R. Etienne-Cummings, T.D. Tran, J. Van der Spiegel, A fully integrated wireless compressed sensing neural signal acquisition system for chronic recording and brain machine interface, IEEE Transactions on Biomedical Circuits and Systems 10 (4) (2016) 874–883.

[24] M. Rezaei, E. Maghsoudloo, C. Bories, Y. De Koninck, B. Gosselin, A low-power current-reuse analog front-end for high-density neural recording implants, IEEE Transactions on Biomedical Circuits and Systems 12 (2) (2018) 271–280.

[25] R.F. Yazicioglu, P. Merken, R. Puers, C. van Hoof, A 200 μW eight-channel EEG acquisition ASIC for ambulatory EEG systems, IEEE Journal of Solid-State Circuits 43 (12) (2008) 3025–3038.

[26] J. Xu, R.F. Yazicioglu, B. Grundlehner, P. Harpe, K.A.A. Makinwa, C. Van Hoof, A 160 μW 8-channel active electrode system for EEG monitoring, IEEE Transactions on Biomedical Circuits and Systems 5 (6) (2011) 555–567.

[27] K.A. Ng, P.K. Chan, A CMOS analog front-end IC for portable EEG/ECG monitoring applications, IEEE Transactions on Circuits and Systems I-Regular Papers 52 (11) (2005) 2335–2347.

[28] H. Wu, Y.P. Xu, A 1V 2.3 μW biomedical signal acquisition IC, in: 2006 IEEE International Solid-State Circuits Conference Digest of Technical Papers, 2006, pp. 119–128.

[29] T. Denison, K. Consoer, W. Santa, A.T. Avestruz, J. Cooley, A. Kelly, A 2 μW 100 nV/rtHz chopper-stabilized instrumentation amplifier for chronic measurement of neural field potentials, IEEE Journal of Solid-State Circuits 42 (12) (2007) 2934–2945.

[30] R.F. Yazicioglu, P. Merken, R. Puers, C. Van Hoof, A 60μW 60 nV/rtHz readout front-end for portable biopotential acquisition systems, IEEE Journal of Solid-State Circuits 42 (5) (2007) 1100–1110.

[31] R. Wu, K.A.A. Makinwa, J.H. Huijsing, A chopper current-feedback instrumentation amplifier with a 1 mHz $1/f$ noise corner and an AC-coupled ripple reduction loop, IEEE Journal of Solid-State Circuits 44 (12) (2009) 3232–3243.

[32] X. Zou, X. Xu, L. Yao, Y. Lian, A 1-V 450-nW fully integrated programmable biomedical sensor interface chip, IEEE Journal of Solid-State Circuits 44 (4) (2009) 1067–1077.

[33] N. Verma, A. Shoeb, J. Bohorquez, J. Dawson, J. Guttag, A.P. Chandrakasan, A micro-power EEG acquisition SoC with integrated feature extraction processor for a chronic seizure detection system, IEEE Journal of Solid-State Circuits 45 (4) (2010) 804–816.

[34] Q.W. Fan, F. Sebastiano, J.H. Huijsing, K.A.A. Makinwa, A 1.8 μW 60 nV/rtHz capacitively-coupled chopper instrumentation amplifier in 65 nm CMOS for wireless sensor nodes, IEEE Journal of Solid-State Circuits 46 (7) (2011) 1534–1543.

[35] F. Zhang, J. Holleman, B.P. Otis, Design of ultra-low power biopotential amplifiers for biosignal acquisition applications, IEEE Transactions on Biomedical Circuits and Systems 6 (4) (2012) 344–355.

[36] Y. Tseng, Y.C. Ho, S.T. Kao, C.C. Su, A 0.09 μW low power front-end biopotential amplifier for biosignal recording, IEEE Transactions on Biomedical Circuits and Systems 6 (5) (2012) 508–516.

[37] J. Yoo, L. Yan, D. El-Damak, M.A. Bin Altaf, A.H. Shoeb, A.P. Chandrakasan, An 8-channel scalable EEG acquisition SoC with patient-specific seizure classification and recording processor, IEEE Journal of Solid-State Circuits 48 (1) (2013) 214–228.

[38] R. Muller, S. Gambini, J.M. Rabaey, A 0.013 mm^2 5μW DC-coupled neural signal acquisition IC with 0.5 V supply, in: 2011 IEEE International Solid-State Circuits Conference Digest of Technical Papers, 2011, pp. 302–304.

[39] D. Han, Y. Zheng, R. Rajkumar, G. Dawe, M. Je, A 0.45V 100-channel neural-recording IC with sub-μW/channel consumption in 0.18 μm CMOS, in: 2013 IEEE International Solid-State Circuits Conference Digest of Technical Papers, 2013, pp. 290–291.

[40] C.C. Enz, G.C. Temes, Circuit techniques for reducing the effects of op-amp imperfections: autozeroing, correlated double sampling, and chopper stabilization, Proceedings of the IEEE 84 (11) (1996) 1584–1614.

[41] H. Nyquist, Thermal agitation of electric charge in conductors, Physical Review 32 (1) (1928) 110–113.

[42] J.B. Johnson, Thermal agitation of electricity in conductors, Physical Review 32 (1) (1928) 97–109.

[43] Y. Chen, A. Basu, M. Je, A digitally assisted, pseudo-resistor-less amplifier in 65 nm CMOS for neural recording applications, in: Proceedings of 2012 IEEE 55th International Midwest Symposium on Circuits and Systems, 2012, pp. 366–369.

[44] K.C. Hsieh, P.R. Gray, D. Senderowicz, D.G. Messerschmitt, A low-noise chopper-stabilized differential switched-capacitor filtering technique, IEEE Journal of Solid-State Circuits 16 (6) (1981) 708–715.

[45] C.C. Enz, E.A. Vittoz, F. Krummenacher, A CMOS chopper amplifier, IEEE Journal of Solid-State Circuits 22 (3) (1987) 335–342.

[46] C. Menolfi, Q.T. Huang, A low-noise CMOS instrumentation amplifier for thermoelectric infrared detectors, IEEE Journal of Solid-State Circuits 32 (7) (1997) 968–976.

[47] C. Menolfi, Q.T. Huang, A fully integrated, untrimmed CMOS instrumentation amplifier with submicrovolt offset, IEEE Journal of Solid-State Circuits 34 (3) (1999) 415–420.

[48] R. Burt, J. Zhang, A micropower chopper-stabilized operational amplifier using a SC notch filter with synchronous integration inside the continuous-time signal path, IEEE Journal of Solid-State Circuits 41 (12) (2006) 2729–2736.

[49] T. Delbruck, C.A. Mead, Adaptive photoreceptor with wide dynamic range, in: Proceedings of IEEE International Symposium on Circuits and Systems, 1994, pp. 339–342.

[50] J.L. Bohorquez, M. Yip, A.P. Chandrakasan, J.L. Dawson, A biomedical sensor interface with a sinc filter and interference cancellation, IEEE Journal of Solid-State Circuits 46 (4) (2011) 746–756.

[51] F. Zhang, A. Mishra, A.G. Richardson, B. Otis, A low-power ECoG/EEG processing IC with integrated multiband energy extractor, IEEE Transactions on Circuits and Systems I-Regular Papers 58 (9) (2011) 2069–2082.

[52] W. Wattanapanitch, M. Fee, R. Sarpeshkar, An energy-efficient micropower neural recording amplifier, IEEE Transactions on Biomedical Circuits and Systems 1 (2) (2007) 136–147.

[53] R.H. Olsson, D.L. Buhl, A.M. Sirota, G. Buzsaki, K.D. Wise, Band-tunable and multiplexed integrated circuits for simultaneous recording and stimulation with microelectrode arrays, IEEE Transactions on Biomedical Engineering 52 (7) (2005) 1303–1311.

[54] M.S. Chae, Z. Yang, M.R. Yuce, L. Hoang, W.T. Liu, A 128-channel 6 mW wireless neural recording IC with spike feature extraction and UWB transmitter, IEEE Transactions on Neural Systems and Rehabilitation Engineering 17 (4) (2009) 312–321.

[55] M.T. Shiue, K.W. Yao, C.S.A. Gong, Tunable high resistance voltage-controlled pseudo-resistor with wide input voltage swing capability, Electronics Letters 47 (6) (2011) 377–378.

[56] B.B. Winter, J.G. Webster, Reduction of interference due to common-mode voltage in biopotential amplifiers, IEEE Transactions on Biomedical Engineering 30 (1) (1983) 58–62.

[57] B.B. Winter, J.G. Webster, Driven-right-leg circuit-design, IEEE Transactions on Biomedical Engineering 30 (1) (1983) 62–66.

[58] L. Fay, V. Misra, R. Sarpeshkar, A micropower electrocardiogram amplifier, IEEE Transactions on Biomedical Circuits and Systems 3 (5) (2009) 312–320.

[59] M.A. Haberman, E.M. Spinelli, A multichannel EEG acquisition scheme based on single ended amplifiers and digital DRL, IEEE Transactions on Biomedical Circuits and Systems 6 (6) (2012) 614–618.

[60] N. Van Helleputte, S. Kim, H. Kim, J.P. Kim, C. Van Hoof, R.F. Yazicioglu, A 160 μA biopotential acquisition IC with fully integrated IA and motion artifact suppression, IEEE Transactions on Biomedical Circuits and Systems 6 (6) (2012) 552–561.

[61] Y.M. Chi, C. Maier, G. Cauwenberghs, Ultra-high input impedance, low noise integrated amplifier for noncontact biopotential sensing, IEEE Journal on Emerging and Selected Topics in Circuits and Systems 1 (4) (2011) 526–535.

CHAPTER 3

ECoG signal coding and digitization

Contents

3.1. Introduction

High-density multi-channel recording of neural electrophysiological signals such as local field potentials (LFP) inside the brain and the electrocorticogram (ECoG) on the cortical surface is essential to driving advances in neuroscience and neuroengineering, by increasing spatial resolution and dynamic range of brain–machine interfaces for high-throughput brain activity mapping, and of neural prostheses for monitoring and treatment of neurological disorders. Great advances in spatial resolution and coverage of neural recording can be obtained by silicon integration of multi-channel brain–computer interfaces with high-density electrode arrays for electrical recording and stimulation [1], and their extreme miniaturization by encapsulating the electrode array along with coil antenna for wireless power and data telemetry within a single mm-sized silicon chip [2]. Although the miniaturization of neural implants and their modular distribution across the brain towards full-brain coverage in high-resolution brain–machine interfaces offers various system level advantages such as better conformity to cortical curvature and a decrease in incidence of tissue inflammation, astroglial scarring, and cell death [3,4], the extreme form factor and energy constraints raise severe challenges in signal quality of neural recording.

The limited amount of power delivery with an on-chip coil and multi-channel neural recording requires extreme energy efficiency in the design of neural recording without compromising its inherent design requirement, low input-referred noise (IRN) [5,6], while also retaining small form factor in the design [7–9]. Full-duplex neural interfaces for closed-loop neural modulation require simultaneous operation of electrical recording and stimulation. Stimulation artifacts produce rapid and large-amplitude transients in the recorded signals that easily overwhelm the neural response

High-Density Integrated Electrocortical Neural Interfaces
https://doi.org/10.1016/B978-0-12-815115-0.00010-6

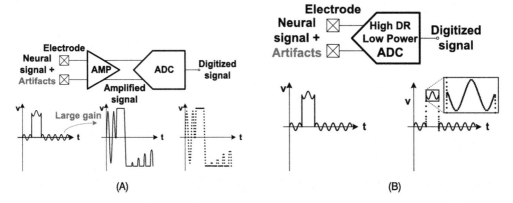

Figure 3.1 Impact of neural recording architecture on dynamic range and transient response: (A) separate analog front-end (AFE) and analog-to-digital converter (ADC) stages; and (B) ADC-direct neural recording.

signals, necessitating a paradigm shift in the design of neural recording toward very high input dynamic range and fast transient response [10].

To resolve small-amplitude neural signals such as LFP and ECoG ranging in the tens of microvolts, typical neural recording circuits include a high-gain, low-noise pre-amplification analog front-end (AFE) stage prior to digitization as illustrated in Fig. 3.1(A) [11]. For low-noise operation the AFE stage typically consumes substantially more power and area than the subsequent analog-to-digital converter (ADC), as the limiting factor in the energy efficiency and integration density of the overall system. Hence most efforts in neural recording design have focused on optimizing the AFE. However, the separation between amplification and digitization stages for neural recording is prone to saturation of the amplified signal under large transients caused by stimulation or motion artifacts. To this end, the latest designs use low-gain (18 dB) pre-amplification in the AFE to mitigate saturation effects [12].

Hybrid architectures utilizing oversampling ADCs with digital feedback to the AFE [8,10] have seen recent adoption due to their increased power and area efficiency. Recent integrated designs combining AFE and ADC in one stage [9] offer further improvements in integration density and expanded input dynamic range. The challenge with previous ADC-direct approaches however is the kT/C sampling noise directly entering the signal path without attenuation, degrading noise–energy efficiency.

To address the confluence of these extreme design challenges for high-density integrated neural recording, a new ADC-direct approach is presented that combines a hybrid analog/digital second order oversampling ADC with predictive digital autoranging (PDA) for high input dynamic range and rapid transient recovery at record noise–energy efficiency [13]. Also kT/C sampling noise is avoided altogether through boxcar sampling [14,15] in mixed-signal feedback [8], while PDA avoids the need for substantial gain attenuation in the feedback loop leading to enhanced signal resolution at higher frequencies. PDA specifically addresses the problem of fast recovery from artifact and stimulation *transients*, by temporarily relaxing resolution through radix-2 expansion of the quantization step size to track large transient slope, and rapidly returning to minimum quantization step noise-limited resolution upon transient completion. Applicable to a wide range of electrophysiological recording applications, the biopotential ADC (BioADC) chip resolves small signals while handling large input transients without saturation as illustrated in Fig. 3.1(B).

The chapter is organized as follows: two widely-used ADC architectures are reviewed in Sect. 3.2 first; the ADC-direct architecture and system operation of neural recording with PDA is described in Sect. 3.3; circuits implementing the architecture are detailed in Sect. 3.4; measurement results are presented in Sect. 3.5; and concluding remarks are offered in Sect. 3.6.

3.2. Widely-used ADC architectures

3.2.1 Successive-approximation ADC

Successive-approximation register (SAR) ADC is the dominant architecture for low-power medium-resolution (8–12 bits) biomedical applications due to its simple architecture involving few analog circuits and its low power consumption at low frequency without static power consumption [16–24]. As shown in Fig. 3.2, SAR ADCs typically consist of input sampling capacitor digital-to-analog converters (DACs), a comparator and SAR logic. Successive approximation of the sampled input performs a binary search over the input range using a time-multiplexing switched-capacitor DAC based on itera-

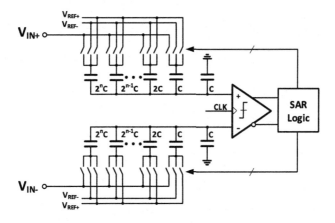

Figure 3.2 A general architecture of differential SAR ADC.

tive partial one-bit quantization results in the order from the most significant bit (MSB) to the least significant bit (LSB) [25,26].

Several techniques and architectures can be applied to minimize power consumption of the SAR ADC. Using a main and sub-binary weighted DAC arrays with a series attenuation capacitor can reduce the total size of the capacitor array, leading to reduced power consumption in the ADC driver and also in the capacitor DAC [27]. Also, the folded capacitor DAC architecture with divided reference voltages reduces the size of capacitor DAC, resulting in further power saving [21]. Using charge-recycling switching methods results in further power savings in the switching of the capacitor DAC [28].

SAR ADCs are generally considered to be most energy efficient for medium-precision low-sampling-rate digitization. However, most micropower SAR ADCs operate at signal levels substantially (3–4 orders of magnitude) greater than typical signal level of physiological signals (Fig. 2.2). They require significant amplification before analog-to-digital conversion for sub-μV resolution. Furthermore, sampling at the Nyquist frequency demands substantially more stringent anti-aliasing analog filtering than required using oversampling techniques. The cost of amplification and anti-aliasing filtering are often not accounted for in ADC energy metrics. Most critically, sampling of biosignals at μV resolution is problematic due to kT/C sampling noise on capacitors which may amount to several tens of μV for typical pF-range capacitors in integrated circuits.

3.2.2 $\Delta\Sigma$ oversampling ADC

$\Delta\Sigma$ ADCs are an alternative solution with the following strengths [29–33]: (1) Resolution and sampling rate can be dynamically reconfigured, with sampling rate proportional to power consumption, so they are adequate for multimodal biopotential sensor applications; (2) They require only few and simple analog components; (3) They are suited for low-power and low-voltage operation; (4) They easily achieve 12–16 bits or higher resolution without complex circuit and layout techniques; (5) Continuous-time $\Delta\Sigma$ topologies are free of kT/C sampling noise and subject to less aliasing and noise folding.

A Gm-C incremental $\Delta\Sigma$ ADC with widely configurable resolution and sampling rate is shown in Fig. 3.3 [29]. A transconductance (G_m) cell converts the differential input voltage signal to a current, approximately linear over the voltage range of typical biopotentials. The difference between this current and a feedback current is integrated and the resulting voltage is compared for three-level quantization of the feedback current, implementing a continuous-time first-order $\Delta\Sigma$ modulator. A continuous-time oversampling ADC avoids the need for anti-aliasing filter and sample-and-hold circuits preceding the ADC. In addition, direct digital control over duty cycle in the feedback offers precise digital gain programmability from 1 to 4096.

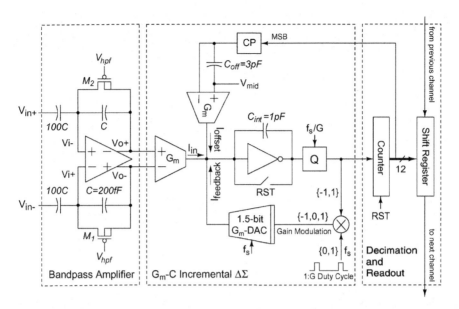

Figure 3.3 An oversampling, aliasing-free biopotential acquisition system utilizing Gm-C incremental $\Delta\Sigma$ ADC [29].

Another example incremental $\Delta\Sigma$ ADC for noninvasive biopotential recording is given in [30]. It receives unbuffered biopotential signals and performs amplification, signal conditioning, and digitization using only a single OTA. Other alternative ADC architectures include a hybrid architecture of SAR and $\Delta\Sigma$ ADC [31], asynchronous level-crossing ADCs [34,35] and a bio-inspired ADC with the successive integrate-and-fire operation [36].

3.3. ADC-direct frontend

Each recording channel features a hybrid analog–digital second-order delta–sigma modulator (2DSM) oversampling ADC, with the biopotential signal coupling directly to the second integrator for high conversion gain and dynamic offset subtraction in the digital domain. More generally as shown in Fig. 3.4(A), with dual inputs u and x into the first and second integrators, respectively, and with additive noise e modeling quantizer error, the dynamics of the 2DSM is given by

$$v[n] = v[n-1] + u[n] - y[n], \tag{3.1}$$

$$w[n+1] = w[n] + v[n] - y[n] + x[n], \tag{3.2}$$

$$y[n] = w[n] + e[n], \tag{3.3}$$

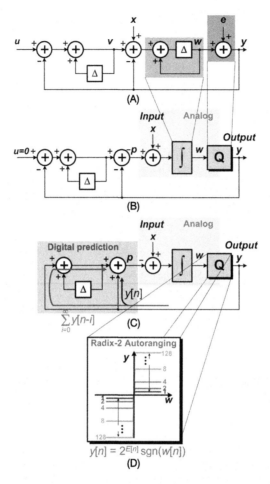

Figure 3.4 ADC-direct neural recording with predictive digital autoranging (PDA). (A) Generalized second-order delta–sigma modulator architecture. (B) Analog continuous-time implementation of the second integrator directly coupled to the input x. (C) Digital accumulator implementation of the first integrator with zero u generating digital prediction p. (D) Predictive digital autoranging (PDA) by radix-2 dynamic expansion and compression in the quantization y of the integrated residue $x - p$.

sampled at discrete time steps $t = nT$. The resulting output

$$
\begin{aligned}
y[n] \;=\; & u[n-1] + \\
& (x[n-1] - x[n-2]) + \\
& (e[n] - 2e[n-1] + e[n-2])
\end{aligned}
\tag{3.4}
$$

produces usual second-order noise shaping with unity gain signal transfer function for an input u, but with first-order differentiation in its signal transfer function for an input x.

Presenting the signal input to the first integrator incurs greater complexity in analog circuit implementation and, more fundamentally, is prone to saturation in the 2DSM loop dynamics, which for $1 - b$ quantization is only conditionally stable for a narrow regime of inputs near zero, $u \approx 0$. In contrast, zeroing the input to the first integrator $u = 0$, and directly coupling the BioADC input x to the second integrator, ensures stable saturation-free 2DSM loop dynamics with only $1 - b$ quantization in the output y.

Continuous-time analog implementation of the second integrator, illustrated in Fig. 3.4(B), obviates the need for sampling the time-varying input $x(t)$. Instead, the integrator continuously integrates the residue between the input $x(t)$ and the piecewise constant digital prediction signal $p[n]$:

$$w[n + 1] = w[n] + \frac{1}{T} \int_{nT}^{(n+1)T} (x(t) - p[n]) \, dt \tag{3.5}$$

where the digital prediction

$$
\begin{aligned}
p[n] &= -v[n] + y[n] \\
&= \sum_{i=0}^{\infty} y[n - i] + y[n] \tag{3.6}
\end{aligned}
$$

is produced by the first integrator digitally accumulating the quantizer feedback, shown in Fig. 3.4(C), efficiently implemented by up/down counting. In turn, the analog input is reconstructed from (3.5) and (3.4) as

$$
\begin{aligned}
x[n] &= \frac{1}{T} \int_{nT}^{(n+1)T} x(t) \, dt \\
&= \sum_{i=0}^{\infty} y[n - i + 1] - (e[n + 1] - e[n]). \tag{3.7}
\end{aligned}
$$

Despite the appearance of first-order (6 dB per octave) noise-shaping of the quantization error $e[n]$ in (3.7), the extra first-order integrating (-6 dB per octave) loop gain in (3.7) contributed by the accumulation in the quantizer output $y[n]$ for digital prediction leads to the same 15 dB increase in dynamic range for every doubling in oversampling ratio (OSR) as with the standard 2DSM [37]. The digital feedback along with the continuous analog integration implements a second-order predictive loop accommodating for potentially large offset and drift at the input, such as electrode DC offset (EDO) in the DC-coupled input, owing to the large loop gain at low frequencies.

The input dynamic range and transient response of the ADC loop are substantially improved by radix-2 autoranging of the quantizer, in which the history of the quantizer bits $D[n] = \text{sgn}(w[n])$ triggers either a factor 2 expansion or contraction in the digital feedback from the quantizer $y[n]$, shown in Fig. 3.4(D):

$$y[n] = 2^{E[n]} D[n] \tag{3.8}$$

where the 3b exponent $E[n] = \{0, 1, \ldots, 7\}$ covers 7 octaves $(1, 2, \ldots, 128)$ in digital gain. A run of five successive decisions with identical polarity increments the exponent expanding the range by two, whereas a run of three alternating polarity decisions decrements the exponent contracting the range by two:

$$
\begin{aligned}
E[n] &\leftarrow E[n-1] + 1 && \text{if} \quad D[n] = \cdots = D[n-4]; \\
E[n] &\leftarrow E[n-1] - 1 && \text{if} \quad D[n] = -D[n-1] = D[n-2]; \\
E[n] &\leftarrow E[n-1] && \text{otherwise.}
\end{aligned}
\tag{3.9}
$$

The rationale for this strategy is that a run of identical polarity decisions $D[n] = D[n-1]\ldots$ indicates the presence of a large transient, necessitating larger steps in digital prediction for faster recovery, whereas a run of alternating polarity decisions $D[n] = -D[n-1]\ldots$ indicates settling within the quantization limit, permitting smaller steps for higher accuracy. The precise choice of lengths of these runs prior to triggering radix-2 expansion/contraction in the range is not critical and is determined through behavioral simulation for optimal overall signal to noise and distortion ratio (SNDR) as a compromise between agility in transient recovery, resilience to variability in the signal, and stability in nonlinear loop dynamics.

The combination of digital prediction and radix-2 autoranging constitutes predictive digital autoranging (PDA). PDA substantially improves transient response to large

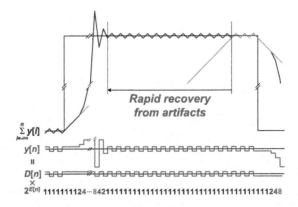

Figure 3.5 Illustration of the effect of autoranging in digital prediction on dynamics of the signal output prior to decimation. Radix-2 predictive digital autoranging (PDA, blue line) (black in print version) significantly improves transient response over fixed-amplitude quantization (grey line) while maintaining same noise-limited LSB-level resolution in steady state.

artifacts, tracking full-swing signal excursions without saturation while quickly reestablishing noise-limited resolution upon settling of the signal as shown in Fig. 3.5. During the artifact transient, PDA temporarily relaxes the resolution with larger quantization step size to accommodate the fast response. This temporary relaxation of ADC resolution during the transient is governed by the following trade-off between transient slope tracking response and quantization step size:

$$2^{E[n]} > \text{maximum signal step size} = \frac{2\pi f_{sig} A_{sig}}{f_{comp}} \tag{3.10}$$

where (for a sinusoidal signal) A_{sig} is the signal amplitude referred to the DAC LSB level (in units of 63 μV referred to the input), f_{sig} is the signal frequency, and f_{comp} is the sampling rate (Fig. 3.6).

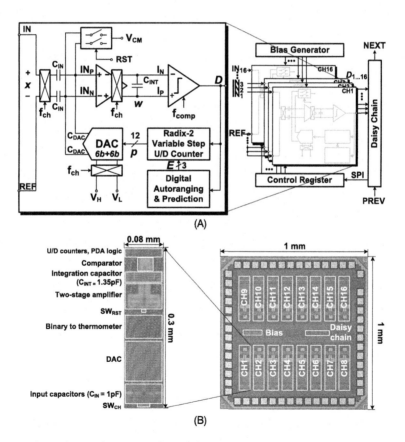

Figure 3.6 16-channel ADC-direct neural recording IC with predictive digital autoranging (PDA). (A) System diagram and circuit architecture with single-channel detail. (B) IC micrograph with corresponding single-channel detail.

3.4. Circuit design implementation

Fig. 3.6 presents the system diagram circuit architecture and micrograph of the 16-channel neural-signal-acquisition integrated circuit (IC), with detailed view of one of 16 identical channels. Each channel implements the PDA hybrid analog–digital 2DSM of Fig. 3.4(D), digitally predicting the analog input at 12-bit resolution from a single-bit quantization of the continuously integrated residue at effective 32 oversampling ratio (OSR).

The continuous differential input $x(t)$ is chopped, and its digital prediction $p[n]$ is reconstituted by a correspondingly reference-chopped 12-bit 6b–6b segmented DAC, prior to constructing the difference through capacitive coupling to the differential inputs IN_P and IN_N of a transconductance amplifier. For low-noise implementation, no specific sampling process through switching of capacitors is utilized and the signal couples to the amplifier input entirely through charge redistribution in capacitive coupling, avoiding kT/C switching noise altogether. The common-mode DC bias at the IN_P and IN_N input nodes is set to V_{CM} by activating two switches at power-on reset, and are subsequently deactivated and remain off throughout the entire operation. Junction diode leakage to bulk connections of these switches towards V_{CM} maintain common-mode DC bias with TΩ-range impedance, with no need for periodic reset.

The resulting residue $x(t) - p[n]$ is transconductance amplified and unchopped to baseband for continuous-time integration onto C_{INT}. A dynamic comparator produces the binary quantizer output $D[n]$, which through barrel-shifting logic is combined with radix-2 autoranging $E[n]$ to produce the quantizer output $y[n]$ consistent with (3.8) and Fig. 3.4(D). The digital prediction $p[n]$ in turn is obtained as the instantaneous sum of the digital feedback $y[n]$ and its running accumulation, completing the second-order loop. The 16 channels on-chip share common reference, bias and control signals, and their outputs $D_{1...16}$ are daisy-chained at the output to enable higher channel counts through cascaded multi-chip configuration. The 16-channel neural recording IC measures 1 mm × 1 mm, with 0.024 mm^2 per channel, in 65 nm low-power bulk CMOS. Realized capacitance values for C_{IN} and C_{INT} are 1 and 1.35 pF, respectively, while the effective capacitance C_{DAC} of the 6b+6b DAC referred to the integrator input is 128 fF.

A 2-stage fully differential amplifier with two independent stages of common-mode feedback shown in Fig. 3.7(A) feeds into an integration capacitor C_{INT}. Current biases for I_{B1} and I_{B2} are set to 375 and 25 nA, respectively. Current-reusing nMOS and pMOS input pairs in the first stage boost transconductance to 22 μS for improved noise efficiency factor (NEF) [38], while 600 mV$_{pp}$ output swing at 0.8 V supply in the second stage increases spurious-free dynamic range. The simulated signal gain of the integrator is greater than 46 dB near the 32 kHz chopping frequency. Auxiliary amplifiers A_{CF} with conventional nMOS input differential pairs implement low-frequency common-mode feedback in each of the two stages, whereas capacitances $C_{CM1} = 15$ fF

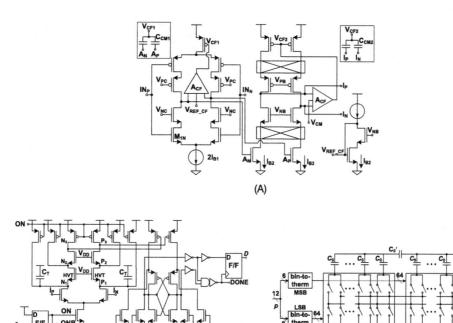

(A)

(B)   (C)

Figure 3.7 Schematics of core analog circuits: (A) transconductance amplifier for continuous-time integrator; (B) dynamic comparator; and (C) 6b+6b segmented capacitive DAC.

and $C_{CM2} = 8$ fF Miller-boosted for common–mode signals stabilize common–mode feedback loops.

A two-stage comparator shown in Fig. 3.7(B) [39] performs 1b quantization. Decision time ranges from 1.5 to 2 μs depending on input amplitude, dominated by capacitive loading ($C_T = 20$ fF) of the first-stage current-starved ($I_C = 20$ nA) preamplifier. Owing to the pre-amplification stage, simulated input-referred noise (INR) of the comparator is less than 80 μV$_{rms}$. At 32 kHz operation, the comparator draws less than 3 nA current from the 0.8 V supply. The ONB clock signal, utilized in subsequent digital logic stages, is asserted when the decision is made.

Each of two differential segmented 6b+6b DACs is implemented with two 64-element custom arrays of 2 fF unit capacitors C_0, bridged by 4% larger capacitor C_0, shown in Fig. 3.7(C). The DAC reference levels V_H and V_L are tied to the supplies $V_{DD} = 0.8$ V and $V_{SS} = 0$ V, respectively. While current consumption from V_H is 50 nA, digital logic within the DAC consumes 10 nA from the 0.8 V supply at 32 kHz.

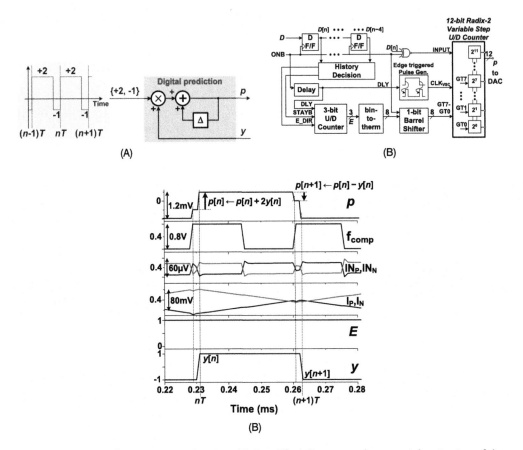

Figure 3.8 PDA implementation and timing. (A) Simplified diagram and sequential activation of the digital prediction stage in Fig. 3.4(C) with up/down counters. (B) Detailed logic implementation. (C) Simulated time-domain waveforms illustrating internal operation.

The implementation and timing control of the PDA is shown in Fig. 3.8. A 12-bit radix-2 variable-step up/down counter implements the update (3.6) in $p[n]$ in two phases: a double increment/decrement step $p[n] \leftarrow p[n] + 2\gamma[n]$ activating the counter at its binary input position $E[n] + 1$, followed by a retracing step with opposite polarity $p[n] \leftarrow p[n] - \gamma[n]$ activating the counter at input position $E[n]$ just before the next cycle. Timing of the two-phase updates in the digital prediction state variable $p[n]$ is triggered by initiation and settling of the comparator output through the ONB signal as shown in the detailed logic diagram in Fig. 3.8(B). The thermometer-coded $(GT0, \ldots, GT7)$ binary input position $E[n]$ of the radix-2 variable-step up/down counter is dynamically adjusted one point higher or lower, or stays put, based on the stored history in the quantization bits $D[n], \ldots, D[n-4]$, according to (3.9). The PDA logic consumes less than 12 nW power at 32 kHz.

3.5. Measurements

Benchtop characterization of several BioADC channels were performed with synthetic data, and *in vivo* validation tests conducted in marmoset primate LFP recording. Unless otherwise noted, $I_S = 1$ µA channel supply current, OSR = 32 oversampling ratio (OSR) and $f_{ch} = 32$ kHz chopping frequency were utilized for all measurements. The input impedance is a function of the chopping frequency and at $f_{ch} = 32$ kHz the measured input impedance is greater than 26 MΩ. The measured common-mode rejection ratio (CMRR), for a 28 mVpp sinusoidal common-mode with zero differential input, is greater than 81 dB from DC to 60 Hz.

Figs. 3.9 and 3.10 show the measured input-referred noise of the BioADC, with input shorted to the reference (IN = REF in Fig. 3.6(A)) Without chopping technique (black line), $1/f$ noise is clearly visible. Chopping above 8 kHz reduces the noise density below 50 nV/$\sqrt{\text{Hz}}$, resulting in 0.99 µV$_{\text{rms}}$ integrated input-referred noise (IRN) over 500 Hz bandwidth and 1.81 noise efficiency factor (NEF) at 32 kHz chopping frequency and 1 µA supply. The major source of this noise is the first stage of the analog

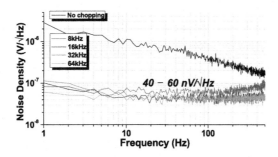

Figure 3.9 Measured input-referred noise spectral density for varying chopping frequency at 1 µA channel current from 0.8 V supply.

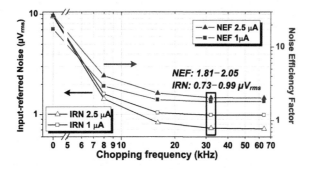

Figure 3.10 Measured integrated input-referred noise for varying chopping frequency and supply current from 0.8 V supply.

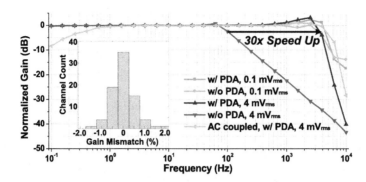

Figure 3.11 Measured large-signal bandwidth, with and without PDA offering a 30× speed improvement. Inset shows gain mismatch across channels.

integrator. Measured IRN across chips is $0.94\ \mu V_{rms}$ on average with $0.1\ \mu V_{rms}$ standard deviation.

The measured effect of PDA on signal-dependent gain is highlighted in Fig. 3.11. Without PDA, the response to a large step transient is slew-rate-limited due to unity increments/decrements in the digital feedback. With PDA, measurements show a 30× speed improvement for $4\ mV_{rms}$ amplitude signals while for small input signals, no significant difference in speed is observed. Indeed, consistent with (3.10), a $4\ mV_{rms}$ signal in the absence of PDA ($E[n] \equiv 0$) starts cutting off for frequencies above 57 Hz at 32 kHz sampling rate, with proportionally higher cut off frequencies at lower signal amplitudes (e.g., 2.3 kHz at $100\ \mu V_{rms}$), whereas activation of PDA achieves full bandwidth limited response independent of signal amplitude by adjusting $E[n] > 0$. The measured 5.95 V/V gain is flat at low frequencies down to DC. Measured relative mismatch (standard deviation over mean) in midband voltage gain is 4.5% across chips (inter-chip) and 0.7% across channels within the same chip (intra-chip).

The measured effect of PDA on increasing input dynamic range is illustrated in Fig. 3.12. PDA extends the input signal range, at greater than 50 dB SNDR, by 22 dB, approaching the full-scale range of the DAC, covering 92 dB input dynamic range. SNDR improvements by PDA at large input signal amplitudes result from both reduced spurs and reduced noise floor, reaching 66 dB at −39 dBV as shown in Fig. 3.12(B)–(C). However, lower than peak SNDR is reached for the larger amplitudes due to nonlinearities in PDA loop dynamics which cause quantization noise and spurs to rise more than proportional to the signal despite the same radix-2 factor simultaneous scaling of both the range and the quantization step by PDA. As such, the rapid transient recovery capability of PDA tracking large slope artifacts comes at a temporary partial loss in signal resolution, which reestablishes its noise-limited level upon completion of the transient. Since typical neural signals are low amplitude and have a $1/f^2$ low-pass power spectrum

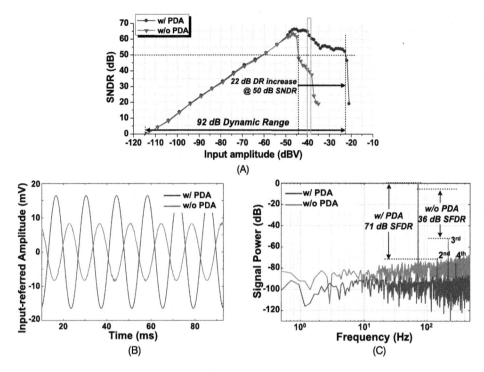

Figure 3.12 Dynamic range with and without PDA. (A) Measured signal-to-noise-and-distortion ratio (SNDR) versus input amplitude. (B) Measured time-domain output, and (C) corresponding spectra showing spurious-free dynamic range (SFDR) at −39 dBV input amplitude.

profile [8], PDA according to (3.10) maintains near-optimal resolution in the absence of artifact transients.

Transient response is significantly improved by PDA. To characterize recovery time in a scenario typical of transients in electrophysiological recording due to movement artifacts or pulsed stimulation, we evaluated PDA response to a synthesized waveform as the combination of two signal sources: one $100\ \mu V_{rms}$ sinusoidal signal and the other a $200\ mV_{PP}$ pulsed artifact transient as shown in Fig. 3.13(A). With PDA, fast tracking in the input was observed, recovering to the $200\ mV_{PP}$ transients in less than 1 ms. In contrast, in absence of PDA the output is markedly slew-limited. The DC-coupled input is capable of capturing slow potentials (≤ 0.1 Hz) while accommodating electrode DC offset (EDO) up to ± 130 mV. For larger EDO, AC-coupled operation is obtained by connecting the DC-coupled input through a pair of external series capacitors (10 nF shown for AC-coupled reference in Fig. 3.11).

Side-by-side comparison between the BioADC and a commercially available benchmark (Intan RHD-2132, [40]) was performed with a combination of synthetic harmonic and transient signals to elicit various metrics in the comparison, shown in

Fig. 3.14. The BioADC consistently demonstrated superior performance in input-referred noise, input dynamic range, and transient response over the RHD-2132 benchmark which shows marked nonlinear transients for even modest (3 mVpp) input transients. The remarkably high high-pass corner of the benchmark in Fig. 3.14(C) is due to signal-dependent nonlinear conductance onto the input node of its AC–coupled front-end amplifier. This default "fast-settling" feature offered in the RHD-2132 for

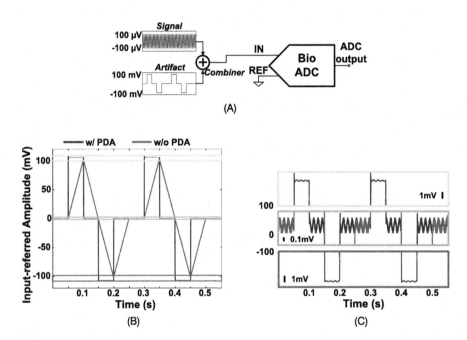

Figure 3.13 Transient response to large artifacts, with and without PDA. (A) Test setup for controlled experiments using synthesized and combined signal and artifacts. (B) Measured time-domain waveforms; and (C) zoom-in waveform showing amplitude detail in settling.

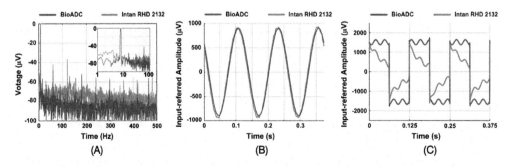

Figure 3.14 Side-by-side comparison of the BioADC with a commercially available neural data acquisition system (Intan RHD-2132 [40], using the default fast-settling feature).

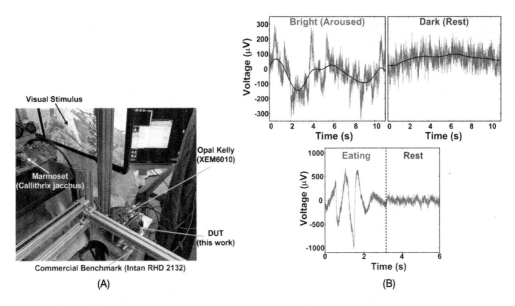

Figure 3.15 (A) Experimental setup for in vivo recording of frontal cortex local field potentials in a marmoset primate subject (Callithrix jacchus) under visual stimulation and (B) *In vivo* LFP recordings. Raw data (blue) (dark grey in print version) is lowpass filtered (≤ 1 Hz) to show slow potentials (black).

rapid recovery from large transients paradoxically introduces slow response transients for a range of amplitudes in signal transients.

In vivo local-field potential (LFP) recordings using the 16-channel neural acquisition IC connecting to a NeuraLynx microwire electrode array inserted in frontal cortex of a marmoset primate (Callithrix jacchus) are shown in Fig. 3.15(A), resolving slow potentials (≤ 0.1 Hz) of 200 μV_{pp} amplitude comparable to the ECoG signal range indicative of subject arousal state that are often missed by AC coupled commercial neural instrumentation unless with severe degradation in SNR [41].

Comparison of key metrics with the state-of-the-art in neural recording ICs is given in Table 3.1. In addition to NEF, the neural ADC achieves a power efficiency factor (PEF) of 2.6, almost a fourfold improvement among integrated front-end ADCs reported in the literature.

3.6. Conclusions

Along with a review of SAR and $\Delta\Sigma$ ADCs, we presented an ADC-direct alternative to conventional approaches to integrated neural recording that alleviates common problems with amplifier saturation during large artifact transients and substantial signal attenuation of low-frequency biopotentials in AC-coupled operation, with further aggravated effects compounding these two through amplitude-dependence of highpass corner frequency.

Table 3.1 Metric comparison with state-of-the-art.

	JSSC'14 [5]	ISSCC'17 [6]	Intan [40]	JSSC'15 [8]	JSSC'17 [9]	VLSI'17 [10]	ISSCC'18 [12]	This work
Power/Ch (μW)	0.97	2.8	100	2.3	0.63	8	7.3	0.8
Supply (V)	1.8	1.2	3.3	0.5	1.2	1	1.2	0.8
Noise density (nV/$\sqrt{Hz}$)[a]	63	127	–	58	101	71	127	44
NEF	1.77[b]	7.4[b]	–	4.76	2.86	7.8	12.18	1.81
PEF (NEF^2 V$_{DD}$)	5.6[b]	66[b]	–	11.3	9.8	60.8	178	2.6
ENOB (bits)	9.57	–	–	–	11.7	10.2	14	10.7
Input DR (dB)	–	81[c]	–	50[c]	–	90	90	92
EDO range (mV$_{pp}$)	AC-coupled	AC-coupled[d]	AC-coupled	100	rail-to-rail	100	AC-coupled[d]	260
CMRR (dB)	–	>85	82	88	88	–	–[e]	81
Area/Ch (mm^2)	–	0.069[b]	0.5[f]	0.025	0.013	–	0.113	0.024
Process (nm)	180	40	–	65	130	180	40	65

[a] Input-referred noise/$\sqrt{Hz}$, differential configuration.
[b] Front-end amplifier only, excluding ADC.
[c] SNDR = 0 dB estimated from input-referred noise.
[d] DC servo loop implements high-pass cut-off for AC-coupled recording.
[e] Tolerance to 700 mV$_{pp}$ common-mode interference.
[f] Estimated.

The unique 2DSM topology with kT/C-noise free input coupling into the chopped second integrator delivers high energy-noise efficiency while PDA handles electrode DC offsets and recovers from transient artifacts, offering a wide input DR. *In vivo* LFP recordings from marmoset primate frontal cortex demonstrate its unique capabilities.

References

[1] M.M. Maharbiz, R. Muller, E. Alon, J.M. Rabaey, J.M. Carmena, Reliable next-generation cortical interfaces for chronic brain-machine interfaces and neuroscience, Proceedings of the IEEE 105 (1) (Jan. 2017) 73–82.

[2] S. Ha, A. Akinin, J. Park, C. Kim, H. Wang, C. Maier, P.P. Mercier, G. Cauwenberghs, Silicon-integrated high-density electrocortical interfaces, Proceedings of the IEEE 105 (1) (2017) 11–33.

[3] G.C. McConnell, H.D. Rees, A.I. Levey, C.-A. Gutekunst, R.E. Gross, R.V. Bellamkonda, Implanted neural electrodes cause chronic, local inflammation that is correlated with local neurodegeneration, Journal of Neural Engineering 6 (5) (2009).

[4] L. Karumbaiah, T. Saxena, D. Carlson, K. Patil, R. Patkar, E.A. Gaupp, M. Betancur, G.B. Stanley, L. Carin, R.V. Bellamkonda, Relationship between intracortical electrode design and chronic recording function, Biomaterials 34 (33) (2013) 8061–8074.

[5] W.M. Chen, H. Chiueh, T.J. Chen, C.L. Ho, C. Jeng, M.D. Ker, C.Y. Lin, Y.C. Huang, C.W. Chou, T.Y. Fan, M.S. Cheng, Y.L. Hsin, S.F. Liang, Y.L. Wang, F.Z. Shaw, Y.H. Huang, C.H. Yang, C.Y. Wu, A fully integrated 8-channel closed-loop neural-prosthetic CMOS SoC for real-time epileptic seizure control, IEEE Journal of Solid-State Circuits 49 (1) (2014) 232–247.

[6] H. Chandrakumar, D. Markovic, A 2.8 μW 80 mVpp-linear-input-range 1.6GΩ-input impedance bio-signal chopper amplifier tolerant to common-mode interference up to 650 mVpp, in: 2017 IEEE International Solid-State Circuits Conference (ISSCC), Feb. 2017, pp. 448–449.

[7] R. Muller, S. Gambini, J.M. Rabaey, A 0.013 mm^2, 5 μW, DC-coupled neural signal acquisition IC with 0.5 V supply, IEEE Journal of Solid-State Circuits 47 (1) (2012) 232–243.

[8] R. Muller, H.-P. Le, W. Li, P. Ledochowitsch, S. Gambini, T. Bjorninen, A. Koralek, J. Carmena, M. Maharbiz, E. Alon, J. Rabaey, A minimally invasive 64-channel wireless μECoG implant, IEEE Journal of Solid-State Circuits 50 (1) (2015) 344–359.

[9] H. Kassiri, M.T. Salam, M.R. Pazhouhandeh, N. Soltani, J.L.P. Velazquez, P. Carlen, R. Genov, Rail-to-rail-input dual-radio 64-channel closed-loop neurostimulator, IEEE Journal of Solid-State Circuits 52 (11) (2017) 2793–2810.

[10] B.C. Johnson, S. Gambini, I. Izyumin, A. Moin, A. Zhou, G. Alexandrov, S.R. Santacruz, J.M. Rabaey, J.M. Carmena, R. Muller, An implantable 700 μW 64-channel neuromodulation IC for simultaneous recording and stimulation with rapid artifact recovery, in: 2017 Symposium on VLSI Circuits, Jun. 2017, pp. C48–C49.

[11] R.R. Harrison, C. Charles, A low-power low-noise CMOS amplifier for neural recording applications, IEEE Journal of Solid-State Circuits 38 (6) (2003) 958–965.

[12] H. Chandrakumar, D. Markovic, A 15.2-ENOB continuous-time $\Delta\Sigma$ ADC for a 200 mVpp-linear-input-range neural recording front-end, in: 2018 IEEE International Solid-State Circuits Conference – (ISSCC), Feb. 2018, pp. 232–234.

[13] C. Kim, S. Joshi, H. Courellis, J. Wang, C. Miller, G. Cauwenberghs, A 92 dB dynamic range sub-μVrms-noise 0.8 μW/ch neural-recording ADC array with predictive digital autoranging, in: 2018 IEEE International Solid-State Circuits Conference – (ISSCC), Feb. 2018, pp. 470–472.

[14] C.D. Ezekwe, B.E. Boser, A mode-matching $\Sigma\Delta$ closed-loop vibratory gyroscope readout interface with a 0.004°/s/$\sqrt{\mathrm{Hz}}$ noise floor over a 50 Hz band, IEEE Journal of Solid-State Circuits 43 (12) (2008) 3039–3048.

[15] H. Jiang, C. Ligouras, S. Nihtianov, K.A.A. Makinwa, A 4.5 nV/$\sqrt{\mathrm{Hz}}$ capacitively coupled continuous-time Sigma–Delta modulator with an energy-efficient chopping scheme, IEEE Solid-State Circuits Letters 1 (2018).

[16] R.F. Yazicioglu, P. Merken, R. Puers, C. van Hoof, A 200 μW eight-channel EEG acquisition ASIC for ambulatory EEG systems, IEEE Journal of Solid-State Circuits 43 (12) (2008) 3025–3038.

[17] N. Van Helleputte, S. Kim, H. Kim, J.P. Kim, C. Van Hoof, R.F. Yazicioglu, A 160 μA biopotential acquisition IC with fully integrated IA and motion artifact suppression, IEEE Transactions on Biomedical Circuits and Systems 6 (6) (2012) 552–561.

[18] J. Yoo, L. Yan, D. El-Damak, M.A. Bin Altaf, A.H. Shoeb, A.P. Chandrakasan, An 8-channel scalable EEG acquisition SoC with patient-specific seizure classification and recording processor, IEEE Journal of Solid-State Circuits 48 (1) (2013) 214–228.

[19] X. Zou, X. Xu, L. Yao, Y. Lian, A 1-V 450-nW fully integrated programmable biomedical sensor interface chip, IEEE Journal of Solid-State Circuits 44 (4) (2009) 1067–1077.

[20] N. Verma, A.P. Chandrakasan, An ultra low energy 12-bit rate-resolution scalable SAR ADC for wireless sensor nodes, IEEE Journal of Solid-State Circuits 42 (6) (2007) 1196–1205.

[21] L. Yan, J. Yoo, B. Kim, H.J. Yoo, A 0.5-μV_{rms} 12-μW wirelessly powered patch-type healthcare sensor for wearable body sensor network, IEEE Journal of Solid-State Circuits 45 (11) (2010) 2356–2365.

[22] S. Lee, L. Yan, T. Roh, S. Hong, H.J. Yoo, A 75 μW real-time scalable body area network controller and a 25 μW ExG sensor IC for compact sleep monitoring applications, IEEE Journal of Solid-State Circuits 47 (1) (2012) 323–334.

[23] M. Khayatzadeh, X. Zhang, J. Tan, W.S. Liew, Y. Lian, A 0.7-V 17.4-μW 3-lead wireless ECG SoC, IEEE Transactions on Biomedical Circuits and Systems 7 (5) (2013) 583–592.

[24] Y.-J. Min, H.-K. Kim, Y.-R. Kang, G.-S. Kim, J. Park, S.-W. Kim, Design of wavelet-based ECG detector for implantable cardiac pacemakers, IEEE Transactions on Biomedical Circuits and Systems 7 (4) (2013) 426–436.

[25] J.L. Mccreary, P.R. Gray, All-MOS charge redistribution analog-to-digital conversion techniques-Part I, IEEE Journal of Solid-State Circuits 10 (6) (1975) 371–379.

[26] R.E. Suarez, P.R. Gray, D.A. Hodges, All-MOS charge redistribution analog-to-digital conversion techniques-Part II, IEEE Journal of Solid-State Circuits 10 (6) (1975) 379–385.

[27] A. Agnes, E. Bonizzoni, P. Malcovati, F. Maloberti, A 9.4-ENOB 1V 3.8 μW 100kS/s SAR ADC with time-domain comparator, in: 2008 IEEE International Solid-State Circuits Conference Digest of Technical Papers, 2008, pp. 246–610.

[28] B.P. Ginsburg, A.P. Chandrakasan, An energy-efficient charge recycling approach for a SAR converter with capacitive DAC, in: Proceedings of the 2005 IEEE International Symposium on Circuits and Systems, 2005, pp. 184–187.

[29] M. Mollazadeh, K. Murari, G. Cauwenberghs, N. Thakor, Micropower CMOS integrated low-noise amplification, filtering, and digitization of multimodal neuropotentials, IEEE Transactions on Biomedical Circuits and Systems 3 (1) (2009) 1–10.

[30] Y.M. Chi, G. Cauwenberghs, Micropower integrated bioamplifier and auto-ranging ADC for wireless and implantable medical instrumentation, in: 2010 Proceedings of the European Solid-State Circuits Conference, 2010, pp. 334–337.

[31] S. Ha, J. Park, Y.M. Chi, J. Viventi, J. Rogers, G. Cauwenberghs, 85 dB dynamic range 1.2 mW 156 kS/s biopotential recording IC for high-density ECoG flexible active electrode array, in: 2013 Proceedings of the European Solid-State Circuits Conference, 2013, pp. 141–144.

[32] J. Garcia, S. Rodriguez, A. Rusu, A low-power CT incremental 3rd order sigma delta ADC for biosensor applications, IEEE Transactions on Circuits and Systems I-Regular Papers 60 (1) (2013) 25–36.

[33] J.R. Custodio, J. Goes, N. Paulino, J.P. Oliveira, E. Bruun, A 1.2-V 165-μW 0.29-mm^2 multibit sigma–delta ADC for hearing aids using nonlinear DACs and with over 91 dB dynamic-range, IEEE Transactions on Biomedical Circuits and Systems 7 (3) (2013) 376–385.

[34] L. Yongjia, Z. Duan, W.A. Serdijn, A sub-microwatt asynchronous level-crossing ADC for biomedical applications, IEEE Transactions on Biomedical Circuits and Systems 7 (2) (2013) 149–157.

[35] W. Tang, A. Osman, D. Kim, B. Goldstein, C.X. Huang, B. Martini, V.A. Pieribone, E. Culurciello, Continuous time level crossing sampling ADC for bio-potential recording systems, IEEE Transactions on Circuits and Systems I-Regular Papers 60 (6) (2013) 1407–1418.

[36] H.Y. Yang, R. Sarpeshkar, A bio-inspired ultra-energy-efficient analog-to-digital converter for biomedical applications, IEEE Transactions on Circuits and Systems I: Regular Papers 53 (11) (2006) 2349–2356.

[37] S. Pavan, R. Schreier, G. Temes, Understanding Delta–Sigma Data Converters, 2nd ed., Wiley-IEEE Press, 2017.

[38] S. Song, M.J. Rooijakkers, P. Harpe, C. Rabotti, M. Mischi, A.H.M. van Roermund, E. Cantatore, A 430 nW 64 nV/$\sqrt{\text{Hz}}$ current-reuse telescopic amplifier for neural recording applications, in: IEEE Biomedical Circuits and Systems Conference (BioCAS), 2013, pp. 322–325.

[39] W. Kim, H.K. Hong, Y.J. Roh, H.W. Kang, S.I. Hwang, D.S. Jo, D.J. Chang, M.J. Seo, S.T. Ryu, A 0.6 V 12 b 10 MS/s low-noise asynchronous SAR-assisted time-interleaved SAR (SATI-SAR) ADC, IEEE Journal of Solid-State Circuits 51 (8) (2016) 1826–1839.

[40] I. Technologies. Rhd2000-series amplifier, [Online]. Available: http://intantech.com/files/Intan_RHD2000_eval_system.pdf.

[41] J.A. Hartings, T. Watanabe, J.P. Dreier, S. Major, L. Vendelbo, M. Fabricius, Recovery of slow potentials in AC-coupled electrocorticography: application to spreading depolarizations in rat and human cerebral cortex, Journal of Neurophysiology 102 (4) (2009) 2563–2575.

CHAPTER 4

Integrated circuit interfaces for electrocortical stimulation

Contents

4.1. Introduction

Implanted electrical neural stimulators (neuromodulators) are devices that inject electrical current into neural tissue via at least two electrodes to either alter the firing conditions of the underlying neurons, or directly induce action potentials. Such devices have been successfully deployed in a wide assortment of applications ranging from stimulation of cortical or deep brain regions to treat neurological disorders like Parkinson's disease, dystonia and epilepsy [1–5], stimulation of sensory cells to restore sensations like hearing or vision in patients who are profoundly deaf or blind [6–9], stimulation of the spinal cord to treat chronic pain [10,11], and beyond [12]. Next-generation neural stimulators are beginning to increase the spatial resolution of stimulation via higher electrode density designs that will enable more precise therapy across a wide range of applications [13–16], while also exploring new directions such as vagus nerve stimulation for targeted pharmaceutical replacement (i.e., electroceuticals) [17].

In all cases, implanted neural stimulators have two important constraints that must be met for long-term use in humans: (1) devices must be sufficiently small to fit within natural human anatomy without significantly damaging or displacing surrounding tissue; and (2) devices must dissipate low power such that local temperature increases do not exceed safety limits [18]. In addition, most practical devices prefer fully wireless operation to minimize the risk of infection otherwise posed by transcutaneous electrical conduit,

High-Density Integrated Electrocortical Neural Interfaces
https://doi.org/10.1016/B978-0-12-815115-0.00011-8
73

thereby necessitating either embedded battery power, transcutaneous wireless power transmission (WPT), or both [19–26]. Since the overall area or volume of implanted neural stimulators is often dictated by the size of energy storage or power-receiving elements (i.e., the battery or WPT coil), whose size in turn is primarily dominated by the average power consumption of the implant itself, minimizing the power dissipation of the implant can be an impactful way to minimize implant size, tissue heating, or potentially both at the same time. In many prior-art neural stimulators, power dissipation is dominated by the energy consumed per stimulation event [27–32], and thus improving the energy efficiency of stimulation can yield significant device-level power reductions.

Unfortunately, it is conventionally difficult for neural stimulators to be both small and energy efficient at the same time. Conventional constant-current stimulators can be implemented in a small area, yet dissipate substantial power across the current source itself when powered by a DC supply voltage [28,33]. On the contrary, adiabatic stimulators, which slowly ramp the supply voltage up and down to minimize the voltage drop across the current source and recycle charge from electrode and tissue capacitance as described in Sect. 4.2.3, can be more energy efficient, yet typically require large off-chip passives to synthesize the adiabatic voltage waveforms. In addition, conventional neural stimulators have difficulty efficiently generating the large voltages necessary to support constant-current stimulation across large ranges of electrode and tissue impedances, and typically do so in wirelessly-powered systems via a two-step rectification/boosting (and/or regulation) process that introduces cascaded losses. As next-generation neural stimulation devices continue to shrink in size, in some cases via full on-chip integration of all necessary neuroinstrumentation functionality [34,35], inclusion of energy-efficient adiabatic stimulation functionality is necessary in a small, fully-integrated form factor.

This chapter first describes and compares representative stimulation methods highlighting how adiabatic stimulation using dynamic power rails improves energy efficiency, and then presents an adiabatic neural stimulator architecture, depicted in Fig. 4.4, that achieves efficient, single-step waveform synthesis, and is fully-integrated on-chip with no external components necessary. Here, adiabatic waveforms are synthesized directly from an on-chip resonant coil by cascading and folding auxiliary rectification stages according to stimulation voltage needs. A prototype of the design is fabricated in a 0.18 μm silicon-on-insulator (SOI) CMOS process, and measurement results reveal the architecture achieves a large output voltage compliance at high efficiency, all in a single fully-integrated chip. To evaluate the effectiveness of stimulation and benchmark against prior-art, this chapter reviews the theory behind adiabatic stimulation and proposes an energy-efficiency figure of merit.

4.2. Electrical stimulation methodologies

Generation of action potentials in underlying neurons through electrical stimulation is typically achieved by passing a sufficient amount of charge to membranes of neurons.

The energy efficiency of neural stimulation strongly depends on the voltage and current waveform of the generated stimulus waveforms, and the characteristics of the employed electrodes [36–39]. Historically, neural stimulation has been accomplished using either constant-voltage or constant-current waveforms due to their relatively simple implementations and demonstrated clinical effectiveness [40]. However, such approaches can suffer from unpredictable charge delivery and/or significant inefficiencies. This section describes the advantages and disadvantages of conventional and emerging stimulation methodologies.

4.2.1 Constant-voltage stimulation

Clinical stimulation devices, for example, those used in deep brain stimulation, traditionally utilized constant-voltage (voltage-controlled) stimulation due to its simplicity in implementation [41,42]. In constant voltage stimulation, power supplies or DC voltage sources such as V_{DD} and V_{SS} are directly connected to the electrode as shown in Fig. 4.1(A). The only required components on the stimulator side are just voltage sources at desired voltage levels.

The resulting stimulation current is determined by the stimulation voltage divided by the impedance summations of the electrode, electrolyte and tissue. As a result, the

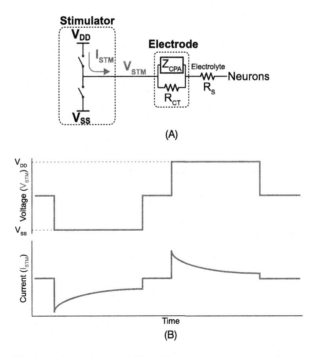

Figure 4.1 (A) Simplified configuration and (B) voltage and current waveforms of constant voltage stimulation.

amount of current delivered to the neuron depended on the impedances based on Ohm's law ($I = V/Z$). Since the impedance of tissue and electrode-tissue interfaces can vary across patients and also with time, the current varies accordingly and its variation is difficult to estimate [43–45].

Furthermore, due to the capacitive component of the electrode impedance, the current has an exponentially decaying waveform as shown in Fig. 4.1(B). Thus, the total delivered charge to the tissue during a stimulation pulse is quite difficult to estimate, and the total charge is difficult to balance, which may lead to generation of toxic byproducts and electrode degradation.

4.2.2 Constant-current stimulation

For a better controllability in charge amount to make sure safety of the tissue and electrode, most recent neural stimulators utilize constant-current (current-controlled) stimulation. By doing so, a constant-current stimulator can provide a set amount of charge to the tissue and evoke desired response regardless of the electrode and tissue impedances, which are unknown and can vary over time [40–42]. As illustrated in Fig. 4.2, a constant current delivered to an electrode results in a ramping electrode

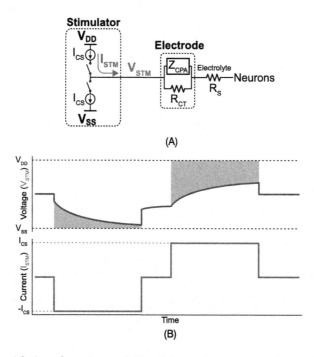

Figure 4.2 (A) Simplified configuration and (B) voltage and current waveforms of constant current stimulation.

voltage. If a high impedance is encountered either in the electrode or within tissue, high compliance (i.e., DC supply) voltages are required for the current source.

In such constant current stimulators, the voltage drop across the current source (i.e., between the DC supply and the ramping electrode voltage), represented by the gray shaded region in Fig. 4.2(B), is energy that is not delivered to tissue, but that is instead dissipated as heat across the current source. As a larger compliance voltage is needed to tolerate a wider range of impedances, the energy wasted across the current source gets much larger. Thus, while constant-current stimulators are generally preferred over constant-voltage stimulators due to increased robustness, they tend to be quite energy inefficient.

4.2.3 Constant-current adiabatic stimulation

To improve the efficiency of constant-current stimulation, it is important to recognize that it is not strictly necessary to operate the stimulating current sources from a fixed supply voltage. An impactful way to reduce unnecessary losses is to operate the current sources from variable supply voltages that closely track the output voltage in order to minimize the voltage drop across the current source. Fig. 4.3 illustrates a nearly-ideal temporal supply voltage for such an adiabatic stimulation arrangement.

Unlike in the case of conventional constant-current stimulation (Fig. 4.2), which draws current during the cathodal phase from a fixed supply rail, V_{SS}, while the electrode voltage, V_{STM}, decreases, here, a variable supply rail, $V_{SS\_Adiabatic}$, is generated such that it dynamically follows the trajectory of the electrode voltage with a small gap, sized to be just large enough to make sure the current source remains in saturation. After the cathodal phase, an anodal phase follows where a constant current is injected from another dynamic supply rail, $V_{DD\_Adiabatic}$. In doing so, $V_{DD\_Adiabatic}$ increases as the electrode voltage increases. By minimizing the voltage gap between the supply voltages and the electrode voltage, energy loss across the current sources can be minimized.

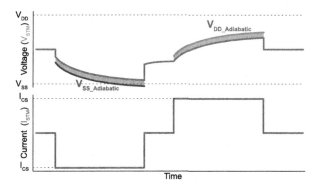

Figure 4.3 Voltage and current waveforms of adiabatic stimulation.

Despite its higher energy efficiency, the adiabatic stimulation method has not been widely deployed in practice, largely due to the complexity and/or size and energy overhead of the circuits needed to create the adiabatic voltage rails. For example, Kelly and Wyatt [28] sequentially switch electrodes between a bank of off-chip capacitors, each biased at increasingly large voltage steps, in order to synthesize a step-wise approximation to a slow voltage ramp. In contrast, Arfin and Sarpeshkar [33] synthesized adiabatic ramp waveforms though a forward-buck/reverse-boost DC–DC converter utilizing an off-chip inductor. As such, despite achieving excellent energy savings, most prior art relied on bulky external components such as capacitors [28] or inductors [33], or alternatively cascaded several power converters, each with substantial loss mechanisms, in order to synthesize the required waveforms [46]. The stimulator in [47] relied on comparison at the RF frequency so that it is limited to applications with low frequency (<10 MHz) RF input.

The following section will present details of the proposed fully integrated stimulator, which synthesizes adiabatic supply rails directly from an on-chip LC tank resonating at 190 MHz, and furthermore recycles the charge stored on the electrode capacitance and delivers it back to V_{DD} in order to further increase energy savings.

4.3. Design details of the stimulator

4.3.1 Overall architecture

The proposed stimulator is integrated into a single-chip neural interfacing system termed ENIAC, an encapsulated neural interfacing and acquisition chip [34,35]. ENIAC includes an on-chip antenna for wireless power and data telemetry, 16 electrodes on the top metal, and all circuitry required for wireless neural recording and stimulation. An illustration of ENIAC, along with a stimulator top-level block diagram, is shown in Fig. 4.4.

The stimulator consists of two major blocks: an adiabatic supply voltage generator, and a constant-current controller. The adiabatic supply voltage generator generates two adiabatic supply voltages, V_{DD_STM} and V_{SS_STM}, directly from the on-chip LC resonant tank. From the two adiabatic supplies, the current controller supplies constant differential currents, I_{STM_U} and I_{STM_L}. The currents flow to the on-chip electrodes through switch multiplexers, which connect the electrode either to the stimulator or to the analog frontend (AFE) used for neural recording. The electrodes, formed by the top-metal layer of the CMOS chip, should be coated with a high-k dielectric material to construct a capacitive electrode, which also enables eliminating any net DC charge transfer. In addition, a capacitive electrode does not have a parallel resistive component, which models a faradaic current. Thus, there is minimal I^2R-based energy dissipation in electrodes while the energy is stored across the electrode by CV^2 instead. The total amount of charge that can be delivered per stimulation phase in ca-

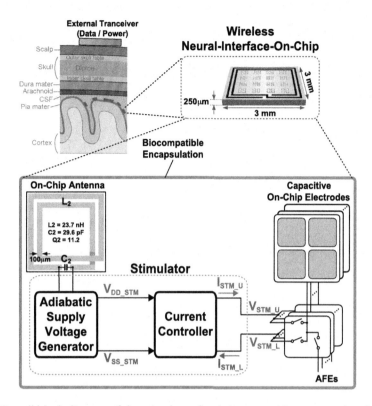

Figure 4.4 Overall block diagram of the stimulator that is integrated in a encapsulated neural interfacing acquisition chip (ENIAC).

pacitive electrodes is proportional to the electrode capacitance and the total voltage rail that the stimulator can generate [35]. While conventional stimulators use high voltage supplies and current sources in a high voltage process, which may lead to poor power and area efficiencies, the proposed stimulator enables both high area- and power-efficiency at high compliance voltage with the strategies described in the following subsections.

4.3.2 Stimulation principle

The operational principle of adiabatic stimulation with ENIAC is described in Fig. 4.5. The stimulation process is composed of two phases: (1) energy provision and (2) energy replenishment. During the first phase, the adiabatic supply voltage generator produces the adiabatic supply voltages, and the differential constant currents flow to differential electrodes as shown on the top side of Fig. 4.5. Since the electrodes are capacitive in ENIAC, the voltages at the electrode V_{STM_U} and V_{STM_L} increase and decrease linearly as the constant stimulation currents flow through them. During this

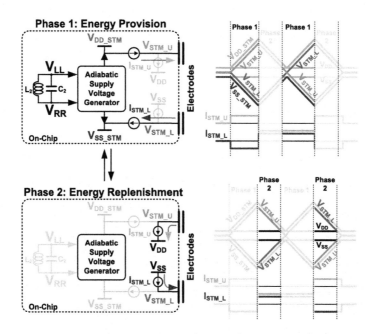

Figure 4.5 Conceptual illustration of the adiabatic stimulation [35].

phase, energy is provided by the stimulator to the tissue through the electrodes. However, a significant amount of the provided energy is accumulated across the capacitive electrodes.

To simplify discussion, a single-ended equivalent model for the proposed adiabatic capacitive stimulation approach, valid over the first phase of stimulation, is illustrated in Fig. 4.6(A). Here, a constant stimulation current, I_{STM}, is provided to the tissue, modeled as resistor R_{TIS}, through the electrode capacitor, C_{EL}. Because I_{STM} is constant over the stimulation time, T_{STM}, the electrode voltage, V_{STM}, increases linearly as shown in Fig. 4.6(B). To not waste energy, the adiabatic supply voltage, V_{DD_STM}, ramps up following V_{STM} with a voltage gap ΔV, which is effectively applied across the current source.

The energy provided by the current source and the energy dissipated and/or stored by the elements can be found by computing the areas denoted in Fig. 4.6(B) because the current is constant over the period of interest. The total energy provided by the current source over time T_{STM} can be calculated as

$$E_{CS} = \frac{1}{2}\{(\Delta V + V_{TIS}) + (V_H + \Delta V + V_{TIS})\} \times I_{STM} \times T_{STM} \qquad (4.1)$$

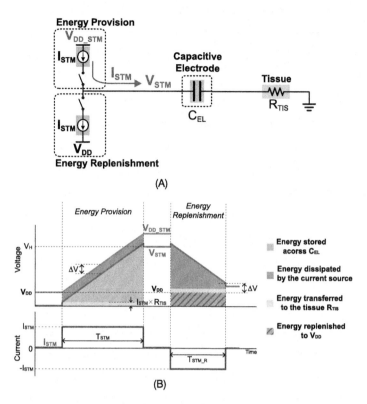

Figure 4.6 Energy usages for adiabatic stimulation. The purple shaded region indicates the energy stored across the capacitive electrode, the gray the dissipated energy at the current source, and the blue the energy delivered to the tissue.

where $V_{TIS} = I_{STM} \cdot R_{TIS}$. Since $I_{STM} \cdot T_{STM} = C_{EL} \cdot V_H$, the above equation can be rewritten as

$$E_{CS} = V_{TIS} I_{STM} T_{STM} + \Delta V I_{STM} T_{STM} + \frac{1}{2} C_{EL} V_H^2. \tag{4.2}$$

Here the first term is the energy transferred to the tissue as highlighted with the blue shadow in Fig. 4.6(B). The second term and the gray shaded area in Fig. 4.6(B) represent the energy dissipated across the current source. The last term is the energy stored across the capacitive electrode corresponding to area of the purple shadow with a triangular shape in Fig. 4.6(B).

As shown, a significantly large part of the provided energy by the stimulation is actually accumulated across the electrode. The energy transferred to the tissue (the first term) is typically much smaller compared to the other two terms. When ΔV is minimized in adiabatic stimulation, almost all provided energy is stored across the capacitor at the end of this stimulation period. For example, assume a realistic case when $I_{STM} = 50 \ \mu A$,

$R_{TIS} = 1$ kΩ, $\Delta V = 0.5$ V, $T_{STM} = 1$ ms, $C_{EL} = 10$ nF, and $V_H = 5$ V. In this case, the stored energy in the capacitive electrode is 125 nJ, the dissipated energy at the current source 25 nJ, and the transferred energy to the tissue 2.5 nJ, while the total energy provided by the stimulator is 152.5 nJ. In this example, 82% of the provided energy is stored across the electrode.

Interestingly, after the first phase of stimulation, conventional stimulators (and, for that matter, conventional digital CMOS logic, which operates on a similar principal) draw the charge down to a negative supply voltage or ground. In doing so, all of the energy stored in the capacitor is discarded to the lowest potential of the stimulator system, and therefore wasted.

To further improve efficiency above the gain introduced by the ramping adiabatic rails, the presented stimulator replenishes the energy nominally stored within the electrode capacitance, and delivers this energy back to the supply rails. The waveforms for energy replenishment as shown in Phase 2 of Fig. 4.5. Here, the stored electrode charge accumulated during the first phase of stimulation is delivered to the power supplies of the system instead of being directly dumped (and thus wasted) to V_{SS}. Here, the capacitive electrode accumulated with the positive voltage, V_{STM_U}, delivers charge to V_{DD}, while the other electrode takes charges from V_{SS}. By doing so, the replenished energy can be reused by other circuits in the system. Since the replenished energy comes back to the system power supplies (V_{DD} and V_{SS}), this contributes to system-level power efficiency.

The various components of energy transfer occurring during replenishment is illustrated in Fig. 4.6. During the energy replenishment phase, charge accumulated in the electrode flows back to the main system supply, V_{DD}. The time period when V_{STM} is larger than $V_{DD} + \Delta V$, denoted by T_{stm_r}, which enables current flow from the electrode to V_{DD}, can be calculated as follows:

$$T_{STM_R} = \frac{V_H + V_{TIS} - V_{DD} - \Delta V}{V_H} \cdot T_{STM}. \qquad (4.3)$$

The replenished energy E_R is

$$E_R = V_{DD} I_{STM} T_{STM_R}. \qquad (4.4)$$

For the example used above, the energy that can be replenished using the scheme is 35.1 nJ, which is 28.4% of the total energy stored in the electrode, assuming $V_{DD} = 1$ V.

4.3.3 Current controller

Redirecting currents between voltage supplies and electrodes is carried out by the current controller shown in Fig. 4.7. During the energy provision phase shown in Fig. 4.7(A), current source $I_{STIM}[4:0]$ is multiplied to I_{STM_L} through a current mirror

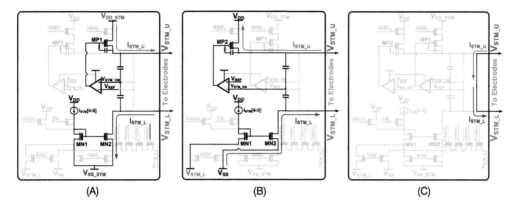

Figure 4.7 Circuit diagrams of the current controller for (A) the energy provision phase, (B) energy replenishment phase, and (C) final phase.

formed by $MN1$ and $MN2$. This I_{STM_L} flows from the electrode to the ramping-down supply voltage V_{SS_STM}. At the same time, an identical current flows from V_{DD_STM} to the other electrode through $MP1$. The gate of $MP1$ is controlled by an amplifier, which makes sure that V_{STM_CM}, the common mode voltage of the electrodes, is equal to a reference voltage, that is, 0.4 V ($\frac{1}{2}V_{DD}$).

During the second phase for energy replenishment, the same current mirror ($MN1$ and $MN2$) used for the previous phase is re-utilized, but now with different power supplies as shown in Fig. 4.7(B). The source of $MN1$ is switched to the electrode voltage V_{STM_L}, while the source and the drain of $MN2$ are swapped in this phase. The lower terminal of $MN2$ serves as the drain during this phase connecting to V_{SS}, while the upper terminal is connected to V_{STM_L} as the source. Here, the source terminals of both $MN1$ and $MN2$ are connected to the common voltage, V_{STM_L}, working as a current mirror. In this manner, current I_{STM_L} flows from V_{SS} to the electrode. Note that $MN2$ has a floating bulk, facilitated by the employed SOI technology.

At the same time during the second phase, an identical current is generated, in this case flowing from the electrode (V_{STM_U}) to V_{DD}. In order to make sure a faster transition and lower peak current during the transition, another pMOSFET $MP2$ and another amplifier are used. Similar to the previous phase, the amplifier maintains the common mode by controlling the current I_{STM_U} to be identical to I_{STM_L}.

A resistor and capacitor are inserted across $MP1$ and $MP2$ for better stability at every phase transition between the energy provision and replenishment. They stabilize the feedback loop formed by the amplifier, $MP1$ and the capacitor between the drain of $MP1$ and the input of the amplifier. The parameter values for the resistor and capacitor are 27.6 kΩ and 1 pF, respectively.

In summary, the accumulated charge at the electrodes are redirected to the system power supplies (V_{DD} and V_{SS}) during the second phase when the electrode voltage are

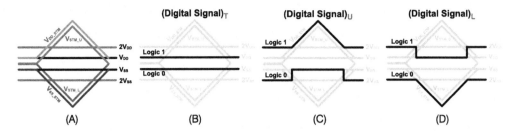

Figure 4.8 (A) Power supply voltages during stimulation. Voltage levels of digital control signals with subscript of (B) T, (C) U and (D) L in Fig. 4.7.

above and below these supplies, respectively. The analog control circuits maintains the stimulation currents to be constant and identical each other while the common-mode voltage remains constant during stimulation.

The first and second phases can be alternated as many times as necessary. In tri-phasic stimulation as depicted in Fig. 4.5, another first phase follows the second phase. At the end any residual charge on the electrodes is then canceled out by shorting as shown in Fig. 4.7(C).

Due to varying adiabatic supply voltages, digital control signals such as *RISE*, *EN*, and *FINISH*, shown in Fig. 4.7, must vary their reference potentials when describing logic 0 and 1 states to be correctly understood and to prevent leakage and gate breakdown. There are three classes of digital signals that must be generated: signals with subscripts T, U, or L. They must be valid within the landscape of possible voltage rails as depicted in Fig. 4.8(A). Signals with subscript T are in the typical voltage range between V_{DD} and V_{SS}, as shown in Fig. 4.8(B), and do not require special consideration. As shown in Fig. 4.8(C), the logic 1 level for signals with subscript U must be adapted to be the higher level between V_{DD_STM} and $2V_{DD}$, which is around 1.6 V in the proposed prototype. The logic 0 level is $2V_{SS}$, which is about -0.8 V, when V_{STM_U} is below 1.2 V, and is switched to V_{SS} when V_{STM_U} increases over 1.2 V. Similarly, the logic levels for signals with subscript L are generated as shown in Fig. 4.8(D).

4.3.4 Adiabatic supply voltage generator

The adiabatic supply voltage generator is shown in Fig. 4.9, and is powered by the on-chip *LC* tank resonating at 190 MHz. The generator produces stimulation power supplies, V_{DD_STM} and V_{SS_STM}, according to the activity occurring with stimulation electrodes V_{STM_U} and V_{STM_L}. Ramping adiabatic supplies, V_{DD_STM} and V_{SS_STM}, are generated by a reconfigurable stack of modified differential Dickson charge pumps starting from V_{DD} and V_{SS}, respectively. The circuits to generate V_{DD_STM} and V_{SS_STM} are symmetric.

The unit cell for the V_{DD_STM} part is shown in Fig. 4.10(A). It includes a pair of cross-coupled nMOSFETs *MN*1 and *MN*2 and a pair of diode-connected pMOSFETs

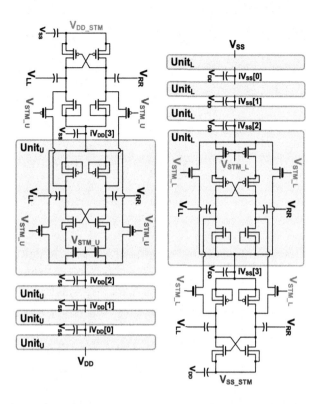

Figure 4.9 Adiabatic supply generator.

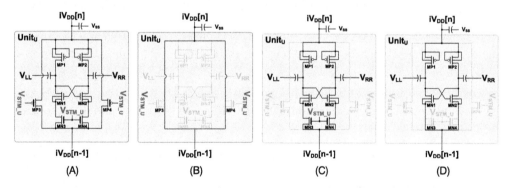

Figure 4.10 (A) Circuit diagram of the unit cell of the adiabatic supply voltage generator. (B)–(D) Three operational phases of the unit cell.

$MP1$ and $MP2$, forming a rectifier. The electrode voltage V_{STM_U} controls a pair of nMOSFETs $MN3$ and $MN4$ on the inner bottom side and two pMOSFETs $MP3$ and $MP4$ on the outer sides.

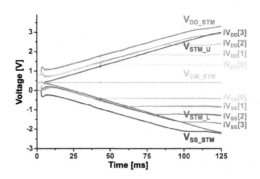

Figure 4.11 Simulation results of the adiabatic supply voltage generator.

Three different operation phases can be distinguished for the unit cell according to V_{STM_U} as shown in Fig. 4.10(B)–(D). When V_{STM_U} is lower than the internal V_{DD}s by more than the threshold voltage of $MP3$ and $MP4$, $MP3$ and $MP4$ form short paths between the upper and lower internal V_{DD}s as shown in Fig. 4.10(B). When V_{STM_U} is in a similar range of the internal V_{DD}s, the outer paths are blocked and the inner part is activated as shown in Fig. 4.10(C). The on-resistances of $MN3$ and $MN4$ are controlled by V_{STM_U} generating a corresponding supply voltage between the upper and lower V_{DD}s. Finally, when V_{STM_U} is higher than the internal V_{DD}, this unit becomes a standard Dickson unit as shown in Fig. 4.10(D).

Simulated waveforms for the adiabatic supply voltage generator are illustrated in Fig. 4.11. As shown, V_{DD_STM} and V_{SS_STM} follow the electrode voltages V_{STM_U} and V_{STM_L} with a small voltage gap. All the while, the common-mode voltage V_{CM_STM} is kept constant. Initially, all the internal stages in the stack are folded near the common-mode voltage. As the electrode voltages V_{STM_U} and V_{STM_L} expand, the first stages become unfolded first and produce $iV_{DD}[0]$ and $iV_{SS}[0]$. After that, the next ones get unfolded one by one until all are unfolded. By doing so, V_{DD_STM} and V_{SS_STM} are generated.

Since the power supplies (V_{DD_STM}, V_{SS_STM}) of the stimulator are directly generated from the LC resonant tank, the overall performance of the circuit, including the maximum stimulation current and highest possible power supply rails is directly related to the LC tank's quality factor and total processed energy. More detailed strategies to achieve a high quality factor and power transfer efficiency in millimeter-sized implants are described in [48,49]. Since the stimulator can draw a large instantaneous current from the LC tank, many power decoupling capacitors were placed both globally and locally near sensitive blocks. A rectifier design that can tolerate rapid variations in the input RF voltage can also help alleviate this issue.

4.4. Experimental validation

The chip was fabricated in a 0.18-μm CMOS silicon-on-insulator (SOI) process, and a die photo of the chip is shown in Fig. 4.12. The total chip dimension is 3×3 mm^2 and has a two-turn on-chip inductor and 16 electrodes of 250×250 μm^2. The core area of the stimulator circuits occupy 0.22 mm^2.

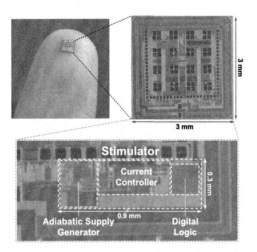

Figure 4.12 Chip micrograph of the stimulator.

Fig. 4.13(A) shows measured 120-μA differential tri-phasic stimulation waveforms and the corresponding currents through RC-modeled electrodes, each composed of a 15 nF capacitor and a 900 Ω resistor in series. As shown in the figure, the two adiabatic supply voltages, V_{DD_STM} and V_{SS_STM}, closely follow the two electrode voltages, V_{STM_U} and V_{STM_L}, respectively, while maintaining a small voltage gap. Voltages V_{DD_STM} and V_{SS_STM} can reach up to -3.3 V and $+3.9$ V, respectively.

Because the current controller described in Sect. 4.3.3 makes sure that the two differential currents (I_{STM_U} and I_{STM_L}) are identical, the differential electrode voltages (V_{STM_U} and V_{STM_L}) should be symmetrical. However, the adiabatic supply voltages are generated by two distinct units within the adiabatic supply voltage generator (described in Sect. 4.3.4), and thus they may be less symmetrical than (V_{STM_U} and V_{STM_L}) as shown in Fig. 4.13(A). Due to this asymmetry, the current source on the V_{SS_STM} side appears to be in the triode region between 1.4 and 1.6 ms, so that the corresponding stimulation current was decreased slightly. A slightly larger RF field delivered to the coil, or a more symmetric design would help to eliminate this issue. As shown on the bottom side of Fig. 4.13(A), the stimulation current waveforms show several peaks, which are generated when the stimulator makes phase changes between the current provision and the energy replenishment phases.

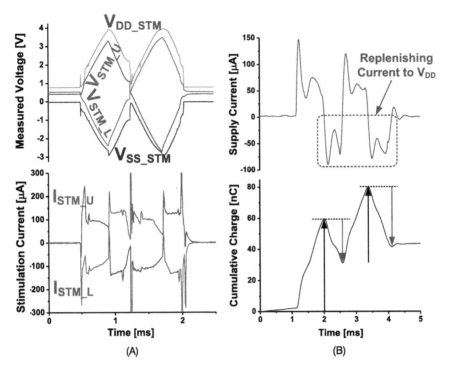

Figure 4.13 (A) (Top) Measured voltage waveforms: adiabatic supply voltages V_{DD_STM} and V_{SS_STM} and electrode voltages V_{STM_U} and V_{STM_L}. (Bottom) The corresponding differential currents. (B) (Top) The current supplied from stimulator V_{DD}. The negative current indicates a replenished current back to the stimulator supply. (Bottom) Accumulated charge of the supply current. The positive ramps denoted with black upward arrows indicate the charge provided from the supply, and the negative ramps denoted with red downward arrows (dark gray in print version) indicate the replenished charge.

The measured supply current from the stimulator is shown in Fig. 4.13(B). Here, negative current indicates energy is being replenished back to V_{DD}. As the computed cumulative charge (Fig. 4.13(B) (Bottom)) shows, more than 63% of charge is returned. As shown in Fig. 4.14, the rectified V_{DD} increases during the energy-replenishing phase, further validating the energy replenishment concept.

The stimulator was further validated across several representative electrode types: 250×250 μm² Pt electrodes in Fig. 4.15(A) and purely resistive electrodes of 50 kΩ in Fig. 4.15(B). As shown in the figure, the stimulator works well not only with capacitive electrodes, but also with other types of electrodes.

In addition, this stimulator supports both tri-phasic and bi-phasic pulse stimulation as shown in Fig. 4.16. Also, it offers individual programmability on each phase's duration and current, and the gap time between the phases.

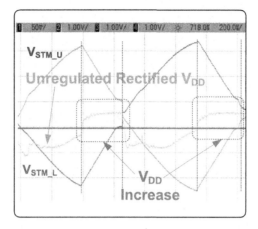

Figure 4.14 A capture of measurement showing V_{DD} increase during the energy replenishing phase. V_{DD} was measured in AC-coupled mode.

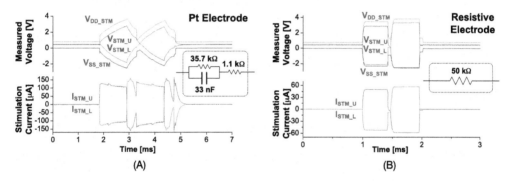

Figure 4.15 Measurements with different electrode models. (A) Pt electrode [50] and (B) purely resistive electrode.

4.5. Benchmarking stimulators

It is challenging to compare different stimulator designs because of different conditions on the operating environments such as electrode size, stimulation current, voltage rails, etc. To help more easily benchmark various stimulation approaches, we propose a new figure of merit (FOM) that quantifies the energy efficiency of stimulation: the Stimulator Efficiency Factor (SEF). This FOM compares the net energy usage of a stimulator compared to one that uses an ideal current source attached to a supply voltage set to the maximum achievable voltage compliance. The net energy usage can be represented as

$$E_{STIM} = (\text{Provided} + \text{Wasted} - \text{Replenished}) \text{ Energy,} \tag{4.5}$$

where the provided energy is the energy that is delivered to the electrode and the tissue, the wasted energy is the energy that is dissipated in the stimulator such as across the

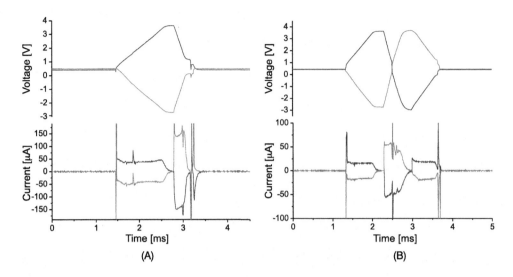

Figure 4.16 Measurement results with (A) bi-phasic and (B) tri-phasic waveforms.

current source, and the replenished energy is the energy that is delivered back to the supply from the energy accumulated in the electrode. Then, *SEF* can be expressed as

$$SEF = \frac{E_{STIM}(\text{Ideal Current Source})}{E_{STIM}(\text{Stimulator})} \tag{4.6}$$

where E_{STIM}(Ideal Current Source) is the net energy usage of a stimulator using a ideal current source and a fixed supply voltage, and E_{STIM}(Stimulator) is of the stimulator under test. Thus, *SEF* represents how much more efficient the stimulator is compared to an ideal current source with DC voltage rails.

Fig. 4.17 illustrates how SEF can be calculated for a stimulator. Here, the presented stimulator is chosen as an example. On the left side in Fig. 4.17, the stimulator's power supplies (V_{DD_STM} and V_{SS_STM}) and electrode voltage (V_{STM_U}) are drawn over a time period of one stimulation event. The shaded area represents the amount of provided (purple), wasted (gray), and replenished (orange) energies, respectively. In order to calculate SEF for this stimulator, a reference stimulator that uses ideal current sources tied to fixed voltages and injects the same amount of current to the same electrode as the stimulator under comparison (i.e., the stimulator that we would like to determine SEF for) must be analyzed. Thus, the electrode voltage of the ideal-current-source stimulator is exactly same as the electrode voltage (V_{STM_U}) of the stimulator under comparison. In addition, the positive supply of the ideal-current-source stimulator (V_{DD_High}) is set as the maximum of the electrode voltage and the negative supply (V_{SS_Low}) as the minimum. Based on these, the net energy usage of the reference ideal-current-source stimulator is calculated, and used for calculating SEF of the stimulator under compari-

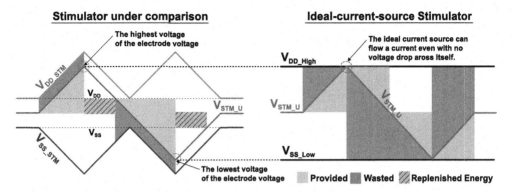

Figure 4.17 Energy delivery, loss and replenishment for the presented adiabatic stimulator and a corresponding stimulator using ideal current sources from constant supply voltages for SEF calculation. V_{DD_High}, the positive supply of the ideal-current-source stimulator, is set to the same voltage as the maximum of the electrode voltage. Likewise, V_{SS_Low}, the negative supply, is set to the same as the minimum of the electrode voltage.

son. By definition, the baseline of an ideal constant current source with DC voltage rails has an SEF of 1. Stimulators with real, non-ideal current sources will obtain an SEF less than 1.

As shown on the right side in Fig. 4.17, the presented stimulator offers further improvement both by the adapting adiabatic rails and by replenishing energy. In order to calculate SEF of the stimulator, the net energy usage was found for the following condition: 120 μA tri-phasic pulse, 1.5 ms duration, and electrodes with 15 nF and 900 Ω in series. The total provided energy was 144 nJ, the wasted energy 18.1 nJ and the replenished energy 27.9 nJ. Resulting E_{STIM} and SEF were 162 nJ and 4.99 without replenishment, respectively. With energy replenishment, E_{STIM} and SEF became 134 nJ and 6.02, respectively. From Eqs. (4.3)–(4.4), the theoretical maximum of the replenished energy E_R is 60.2 nJ. For an ideal case without any energy dissipation or voltage drop, E_{STIM} can be low as 83.8 nJ resulting in SEF of 9.65.

Table 4.1 shows a comparison of state-of-the-art adiabatic stimulators. In comparing this work with other adiabatic stimulators, note that this stimulator is directly powered by a 190 MHz resonant LC tank while other works are powered at frequencies below 10 MHz or from a DC power source. For wireless implantable applications, DC power requires rectification and regulation and additional steps of power conditioning to generate adiabatic power rails, each causing extra energy losses, which are not accounted for in the table. On the contrary, our stimulator produces adiabatic power rails directly from the RF input, over a range that is 9 times larger than V_{DD}. While some other adiabatic stimulators require large external capacitors and inductors, our work requires no external components. This chip achieves more than 60% of charge replenishing ratio, and an SEF of 6.02.

Table 4.1 Comparison of state-of-the-art adiabatic stimulators.

	[28]	[33]	[47]	[46]	This work
Technology [μm]	1.5	0.35	0.18	0.065	0.18 SOI
IC Area [mm²]	4.76[a]	<0.58[b]	0.11[b]	0.45[c]	0.22[b]
External components	LC tank Capacitors, Electrodes	Inductor Electrodes	LC tank Electrodes	Electrode	None
Source of stimulation power	125 kHz External LC	V_{DD}	5 MHz External LC	V_{DD}	190 MHz On-Chip LC
Supply voltage [V]	±1.75	3.3	N/A	1	0.8
Stimulation supply voltage range [V]	±1.75 ($1 \times V_{DD}$)	3.3 ($1 \times V_{DD}$)	N/A	$0 \sim 8.7$ ($8.7 \times V_{DD}$)	$-3.3 \sim +3.9$ ($9 \times V_{DD}$)
Maximum stimulation current [μA]	400	450	140	>500	145
Charge replenishing ratio	N/A	N/A	N/A	N/A	63.1%
Stimulator efficiency factor	$2.13 \sim 2.94$[d]	3.5[d]	$1.33 \sim 5$[d]	N/A	4.99 (w/o replenishment) 6.02[e] (w/ replenishment)
Electrode model	980 nF+1.15 kΩ	930 nF+1 kΩ	5.5 nF//6 kΩ+4 kΩ	10 nF+1 kΩ	4 nF+250 Ω, 15 nF+900 Ω 50 kΩ, 33 nF//36 kΩ+1.1 kΩ

[a] Die area.
[b] Active area.
[c] Active area of the 1:7 DC–DC converter and one channel neural stimulator.
[d] Estimated based on the reported numbers and figures.
[e] Estimated.

4.6. Conclusions

We have presented and demonstrated an adiabatic stimulator that is fully integrated in a mm-sized stand-alone neural interface chip. The proposed stimulator has two distinct advantages. First, the adiabatic stimulation supplies are directly generated from RF while conventional approaches require cascade power conditioning circuits, large external passives, or both, leading to lower overall efficiency. Second, it replenishes energy from the charged electrodes, resulting in more energy-efficient operation. For benchmarking the stimulator performance, a stimulator efficiency factor (SEF) is proposed. This FOM compares the stimulator with an ideal stimulator operating with current sources from constant voltage rails set at the minimum and maximum of the required stimulation voltage compliance. The proposed adiabatic stimulator achieves a measured SEF of 5 excluding replenished energy, and SEF of 6 with estimated replenished energy.

References

[1] A. Fasano, A. Daniele, A. Albanese, Treatment of motor and non-motor features of Parkinson's disease with deep brain stimulation, The Lancet Neurology 11 (5) (2012) 429–442.

[2] L. Vercueil, P. Pollak, V. Fraix, E. Caputo, E. Moro, A. Benazzouz, J. Xie, A. Koudsie, A.-L. Benabid, Deep brain stimulation in the treatment of severe dystonia, Journal of Neurology 248 (8) (2001) 695–700.

[3] A. Berényi, M. Belluscio, D. Mao, G. Buzsáki, Closed-loop control of epilepsy by transcranial electrical stimulation, Science 337 (6095) (2012) 735–737.

[4] D.J. Mogul, W. van Drongelen, Electrical control of epilepsy, Annual Review of Biomedical Engineering 16 (2014) 483–504.

[5] C.O. Oluigbo, A. Salma, A.R. Rezai, Deep brain stimulation for neurological disorders, IEEE Reviews in Biomedical Engineering 5 (2012) 88–99.

[6] J.D. Weiland, M.S. Humayun, Retinal prosthesis, IEEE Transactions on Biomedical Engineering 61 (5) (2014) 1412–1424.

[7] A.C. Ho, M.S. Humayun, J.D. Dorn, L. da Cruz, G. Dagnelie, J. Handa, P.O. Barale, J.A. Sahel, P.E. Stanga, F. Hafezi, A.B. Safran, J. Salzmann, A. Santos, D. Birch, R. Spencer, A.V. Cideciyan, E. de Juan, J.L. Duncan, D. Eliott, A. Fawzi, L.C. Olmos de Koo, G.C. Brown, J.A. Haller, C.D. Regillo, L.V. Del Priore, A. Arditi, D.R. Geruschat, R.J. Greenberg, Long-term results from an epiretinal prosthesis to restore sight to the blind, Ophthalmology 122 (8) (2015) 1547–1554.

[8] K. Stingl, K.U. Bartz-Schmidt, D. Besch, C.K. Chee, C.L. Cottriall, F. Gekeler, M. Groppe, T.L. Jackson, R.E. MacLaren, A. Koitschev, A. Kusnyerik, J. Neffendorf, J. Nemeth, M.A.N. Naeem, T. Peters, J.D. Ramsden, H. Sachs, A. Simpson, M.S. Singh, B. Wilhelm, D. Wong, E. Zrenner, Sub-retinal visual implant Alpha IMS – clinical trial interim report, Vision Research 111 (2015) 149–160, Part B.

[9] S. Ha, M.L. Khraiche, A. Akinin, Y. Jing, S. Damle, Y. Kuang, S. Bauchner, Y.-H. Lo, W.R. Freeman, G.A. Silva, G. Cauwenberghs, Towards high-resolution retinal prostheses with direct optical addressing and inductive telemetry, Journal of Neural Engineering 13 (5) (2016) 056008.

[10] R.B. North, Neural interface devices: spinal cord stimulation technology, Proceedings of the IEEE 96 (7) (2008) 1108–1119.

[11] L. Mazzola, J. Isnard, R. Peyron, F. Mauguière, Stimulation of the human cortex and the experience of pain: Wilder Penfield's observations revisited, Brain 135 (2) (2012) 631–640.

[12] T.W. Berger, G. Gerhardt, M.A. Liker, W. Soussou, The impact of neurotechnology on rehabilitation, IEEE Reviews in Biomedical Engineering 1 (2008) 157–197.

[13] R. Shulyzki, K. Abdelhalim, A. Bagheri, M.T. Salam, C.M. Florez, J.L.P. Velazquez, P.L. Carlen, R. Genov, 320-channel active probe for high-resolution neuromonitoring and responsive neurostimulation, IEEE Transactions on Biomedical Circuits and Systems 9 (1) (2015) 34–49.

[14] H. Kassiri, M.T. Salam, M.R. Pazhouhandeh, N. Soltani, J.L.P. Velazquez, P. Carlen, R. Genov, Rail-to-rail-input dual-radio 64-channel closed-loop neurostimulator, IEEE Journal of Solid-State Circuits 52 (11) (2017) 2793–2810.

[15] K. Mathieson, J. Loudin, G. Goetz, P. Huie, L.L. Wang, T.I. Kamins, L. Galambos, R. Smith, J.S. Harris, A. Sher, D. Palanker, Photovoltaic retinal prosthesis with high pixel density, Nature Photonics 6 (6) (2012) 391–397.

[16] J. Dragas, V. Viswam, A. Shadmani, Y. Chen, R. Bounik, A. Stettler, M. Radivojevic, S. Geissler, M.E.J. Obien, J. Müller, A. Hierlemann, In Vitro multi-functional microelectrode array featuring 59 760 electrodes, 2048 electrophysiology channels, stimulation, impedance measurement, and neurotransmitter detection channels, IEEE Journal of Solid-State Circuits 52 (6) (2017) 1576–1590.

[17] E. Ben-Menachem, D. Revesz, B.J. Simon, S. Silberstein, Surgically implanted and non-invasive vagus nerve stimulation: a review of efficacy, safety and tolerability, European Journal of Neurology 22 (9) (2015) 1260–1268.

[18] P.D. Wolf, Thermal Considerations for the Design of an Implanted Cortical Brain-Machine Interface (BMI), Frontiers in Neuroengineering, Boca Raton (FL), 2008.

[19] M. Mark, T. Bjorninen, Y.D. Chen, S. Venkatraman, L. Ukkonen, L. Sydanheimo, J.M. Carmena, J.M. Rabaey, Wireless channel characterization for mm-size neural implants, in: Proceedings of the Annual International Conference of the IEEE Engineering in Medicine and Biology Society (EMBC), 2010, pp. 1565–1568.

[20] M. Rasouli, L.S.J. Phee, Energy sources and their development for application in medical devices, Expert Review of Medical Devices 7 (5) (2010) 693–709.

[21] J. Walk, J. Weber, C. Soell, R. Weigel, G. Fischer, T. Ussmueller, Remote powered medical implants for telemonitoring, Proceedings of the IEEE 102 (11) (2014) 1811–1832.

[22] A. Kim, M. Ochoa, R. Rahimi, B. Ziaie, New and emerging energy sources for implantable wireless microdevices, IEEE Access 3 (2015) 89–98.

[23] A. Abid, J.M. O'Brien, T. Bensel, C. Cleveland, L. Booth, B.R. Smith, R. Langer, G. Traverso, Wireless power transfer to millimeter-sized gastrointestinal electronics validated in a swine model, Scientific Reports 7 (2017).

[24] Y. Tanabe, J.S. Ho, J.Y. Liu, S.Y. Liao, Z. Zhen, S. Hsu, C. Shuto, Z.Y. Zhu, A. Ma, C. Vassos, P. Chen, H.F. Tse, A.S.Y. Poon, High-performance wireless powering for peripheral nerve neuromodulation systems, Plos One 12 (10) (2017).

[25] D.R. Agrawal, Y. Tanabe, D.S. Weng, A. Ma, S. Hsu, S.Y. Liao, Z. Zhen, Z.Y. Zhu, C.B.W. Sun, Z.Y. Dong, F.Y. Yang, H.F. Tse, A.S.Y. Poon, J.S. Ho, Conformal phased surfaces for wireless powering of bioelectronic microdevices, Nature Biomedical Engineering 1 (3) (2017).

[26] T.P. Delhaye, N. Andre, S. Gilet, C. Gimeno, L.A. Francis, D. Flandre, High-efficiency wireless power transfer for mm-size biomedical implants, IEEE Sensors (2017) 654–656.

[27] C. Kuanfu, Z. Yang, H. Linh, J. Weiland, M. Humayun, L. Wentai, An integrated 256-channel epiretinal prosthesis, IEEE Journal of Solid-State Circuits 45 (9) (2010) 1946–1956.

[28] S.K. Kelly, J.L. Wyatt, A power-efficient neural tissue stimulator with energy recovery, IEEE Transactions on Biomedical Circuits and Systems 5 (1) (2011) 20–29.

[29] E. Noorsal, K. Sooksood, X. Hongcheng, R. Hornig, J. Becker, M. Ortmanns, A neural stimulator frontend with high-voltage compliance and programmable pulse shape for epiretinal implants, IEEE Journal of Solid-State Circuits 47 (1) (2012) 244–256.

[30] M. Monge, M. Raj, M.H. Nazari, C. Han-Chieh, Z. Yu, J.D. Weiland, M.S. Humayun, T. Yu-Chong, A. Emami, A fully intraocular high-density self-calibrating epiretinal prosthesis, IEEE Transactions on Biomedical Circuits and Systems 7 (6) (2013) 747–760.

[31] M. Yip, J. Rui, H.H. Nakajima, K.M. Stankovic, A.P. Chandrakasan, A fully-implantable cochlear implant soc with piezoelectric middle-ear sensor and arbitrary waveform neural stimulation, IEEE Journal of Solid-State Circuits 50 (1) (2015) 214–229.

[32] H.-M. Lee, K.Y. Kwon, L. Wen, M. Ghovanloo, A power-efficient switched-capacitor stimulating system for electrical/optical deep brain stimulation, IEEE Journal of Solid-State Circuits 50 (1) (2015) 360–374.

[33] S.K. Arfin, R. Sarpeshkar, An energy-efficient, adiabatic electrode stimulator with inductive energy recycling and feedback current regulation, IEEE Transactions on Biomedical Circuits and Systems 6 (1) (2012) 1–14.

[34] S. Ha, A. Akinin, J. Park, C. Kim, H. Wang, C. Maier, G. Cauwenberghs, P.P. Mercier, A 16-channel wireless neural interfacing SoC with RF-powered energy-replenishing adiabatic stimulation, in: Symposium on VLSI Circuits Digest of Technical Papers, 2015, pp. C106–C107.

[35] S. Ha, A. Akinin, J. Park, C. Kim, H. Wang, C. Maier, P.P. Mercier, G. Cauwenberghs, Silicon-integrated high-density electrocortical interfaces, Proceedings of the IEEE 105 (1) (2017) 11–33.

[36] M. Sahin, Y. Tie, Non-rectangular waveforms for neural stimulation with practical electrodes, Journal of Neural Engineering 4 (3) (2007) 227–233.

[37] A. Wongsarnpigoon, J.P. Woock, W.M. Grill, Efficiency analysis of waveform shape for electrical excitation of nerve fibers, IEEE Transactions on Neural Systems and Rehabilitation Engineering 18 (3) (2010) 319–328.

[38] T.J. Foutz, D.M. Ackermann Jr, K.L. Kilgore, C.C. McIntyre, Energy efficient neural stimulation: coupling circuit design and membrane biophysics, Plos One 7 (12) (2012) e51901.

[39] H.M. Lee, B. Howell, W.M. Grill, M. Ghovanloo, Stimulation efficiency with decaying exponential waveforms in a wirelessly-powered switched-capacitor discharge stimulation system, IEEE Transactions on Biomedical Engineering 65 (5) (2018) 1095–1106.

[40] F.R. de Noriega, R. Eitan, O. Marmor, A. Lavi, E. Linetzky, H. Bergman, Z. Israel, Constant current versus constant voltage subthalamic nucleus deep brain stimulation in Parkinson's disease, Stereotactic and Functional Neurosurgery 93 (2) (2015) 114–121.

[41] F. Preda, C. Cavandoli, C. Lettieri, M. Pilleri, A. Antonini, R. Eleopra, M. Mondani, A. Martinuzzi, S. Sarubbo, G. Ghisellini, A. Trezza, M.A. Cavallo, A. Landi, M. Sensi, Switching from constant voltage to constant current in deep brain stimulation: a multicenter experience of mixed implants for movement disorders, European Journal of Neurology 23 (1) (2016) 190–195.

[42] P. Amami, M.M. Mascia, A. Franzini, F. Saba, A. Albanese, Shifting from constant-voltage to constant-current in Parkinson's disease patients with chronic stimulation, Neurological Sciences 38 (8) (2017) 1505–1508.

[43] S.F. Lempka, M.D. Johnson, S. Miocinovic, J.L. Vitek, C.C. McIntyre, Current-controlled deep brain stimulation reduces in vivo voltage fluctuations observed during voltage-controlled stimulation, Clinical Neurophysiology 121 (12) (2010) 2128–2133.

[44] T. Cheung, M. Nuno, M. Hoffman, M. Katz, C. Kilbane, R. Alterman, M. Tagliati, Longitudinal impedance variability in patients with chronically implanted DBS devices, Brain Stimulation 6 (5) (2013) 746–751.

[45] K.A. Sillay, J.C. Chen, E.B. Montgomery, Long-term measurement of therapeutic electrode impedance in deep brain stimulation, Neuromodulation 13 (3) (2010) 195–200.

[46] W. Biederman, D.J. Yeager, N. Narevsky, J. Leverett, R. Neely, J.M. Carmena, E. Alon, J.M. Rabaey, A 4.78 mm^2 fully-integrated neuromodulation soc combining 64 acquisition channels with digital compression and simultaneous dual stimulation, IEEE Journal of Solid-State Circuits 50 (4) (2015) 1038–1047.

[47] U. Çilingiroğlu, S. İpek, A zero-voltage switching technique for minimizing the current-source power of implanted stimulators, IEEE Transactions on Biomedical Circuits and Systems 7 (4) (2013) 469–479.

[48] C. Kim, S. Ha, A. Akinin, J. Park, R. Kubendran, H. Wang, P.P. Mercier, G. Cauwenberghs, Design of miniaturized wireless power receivers for mm-sized implants, in: Proceedings of the 2017 IEEE Custom Integrated Circuits Conference (CICC).

[49] J. Park, C. Kim, A. Akinin, S. Ha, G. Cauwenberghs, P.P. Mercier, Wireless powering of mm-scale fully-on-chip neural interfaces, in: Proceedings of the 2017 IEEE Biomedical Circuits and Systems Conference (BioCAS).

[50] W. Franks, I. Schenker, P. Schmutz, A. Hierlemann, Impedance characterization and modeling of electrodes for biomedical applications, IEEE Transactions on Biomedical Engineering 52 (7) (2005) 1295–1302.

CHAPTER 5

Power management for mm-sized ECoG implants

Contents

5.1. Introduction

Miniaturized, highly energy-efficient wireless power transfer (WPT) integrated circuits are critical components in the development of fully-encapsulated implanted brain–computer interface systems that will, through additional advances in improved spatial resolution and coverage of neural recording and stimulation electrodes, enable next-generation neuroscience and neurology experimentation. As an example, modular mm-sized wireless implants which lie directly on the cortical surface, as illustrated in Fig. 5.1, can support more accurate recording of local brainwaves and a higher spatial resolution (≤ 1 mm) than conventional electrocorticography (ECoG) approaches [1,2]. In addition to spatial coverage and density benefits, recent studies have suggested that small devices can reduce incidence of tissue inflammation, astroglial scarring, and cell death [3–6], thereby underscoring the need to miniaturize implants.

Since the size of miniaturized implanted devices is often limited by the size of the embedded WPT coil and corresponding power management and energy-storage circuitry, significant research efforts have been directed at reducing the size of such components. For example, it has been shown that increasing the RF carrier frequency

High-Density Integrated Electrocortical Neural Interfaces
https://doi.org/10.1016/B978-0-12-815115-0.00012-X

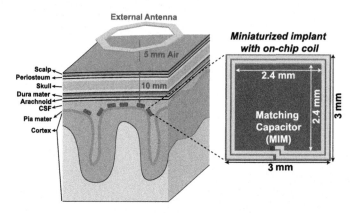

Figure 5.1 Transcutaneous wireless power delivery to mm-sized distributed implants on the cortical surface, each with an integrated power receiving coil.

used for WPT can yield a reduction in receiver size [7–11], since a higher resonant frequency: (i) increases induction via a higher rate of incident magnetic flux over a small receiving coil [12,13], (ii) increases the quality factor (Q) of both power transmitting (TX) and receiving (RX) coils [6,12], and (iii) reduces the area required of the resonant matching capacitor. On the other hand, tissue absorption of electromagnetic waves increases with frequency, resulting in additional tissue losses that ultimately limit the efficiency of WPT. Making matters worse, increased tissue absorption also limits the amount of allowable transmission power due to regulations on the specific absorption rate (SAR) of RF power in human tissue. Balancing of these considerations leads to optimal resonant frequencies in the 100 MHz–1 GHz range under various conditions, enabling the design of mm-sized implants that can efficiently receive sufficient power to operate load circuits under SAR constraints [12–14,6,15].

The pursuit of implant miniaturization also requires research on miniaturizing the rest of the power management and energy storage circuitry. For example, most conventional wirelessly-powered implants utilize a segmented architecture, where the WPT coil is physically separated from the rectifying and regulating circuits. Such an approach requires complex interconnect, packaging, and discrete components that occupy unwanted volume and increase cost [12].

To address these issues and enable further miniaturization of wirelessly-powered implants, this work presents the design of a fully-integrated resonant regulating rectifier (IR^3) that performs voltage and power conversion from an integrated 3 mm × 3 mm on-chip coil to loads; no external components are required for operation. By combining rectification and regulation into a single stage, as illustrated in Fig. 5.2(B), the proposed IR^3 eliminates the conventionally-required inter-stage decoupling capacitance, saving significant volume and eliminating cascaded losses for high efficiency. A hybrid pulse

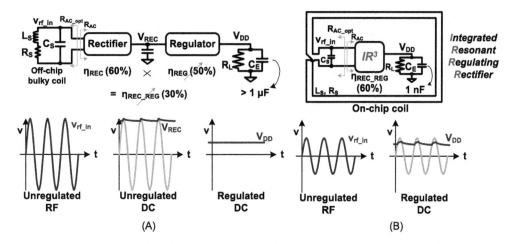

Figure 5.2 Rectification and regulation methodology comparison: (A) conventional separate rectification and regulation conversion; (B) integration of rectification and regulation for one-step conversion without cascaded loss in power conversion efficiency (PCE) and voltage conversion efficiency (VCE).

modulation control scheme is also proposed to enable power efficient rectification and regulation over a large range of RF input amplitude and load current.

Simplified circuit schematics and initial measurement results of the IR^3 were presented in [16]. This chapter presents analysis and optimization of system-level parameters, significantly more detailed circuit schematics, and measurements and characterization of IR^3 performance under varying input and load conditions. This chapter is organized as follows: limitations of conventional segmented architectures motivating the integration of resonant rectification and regulation are discussed in Sect. 5.2. The proposed operation of IR^3 with hybrid pulse frequency and width modulation is outlined in Sect. 5.3, with circuit details elaborated in Sect. 5.4, and simulated and measured results presented in Sect. 5.6. Finally, conclusions are given in Sect. 5.7.

5.2. Architectural considerations

The purpose of this section is to show that performing combined rectification and regulation with an integrated coil is not only advantageous from a size perspective, but that merging these two functions into a single stage has tangible system-level efficiency benefits. To demonstrate this, we first show that to maximize overall system efficiency in mm-sized systems, it is necessary to maximize not only PCE, but also VCE. Then, we show how maximizing both VCE and PCE in conventional cascaded systems is difficult, while attaining high efficiency can be more easily achieved by merging rectification and regulation.

5.2.1 Importance of voltage and power conversion efficiency

The wireless power transfer system efficiency (WSE, the net power gain from the transmitted power at the TX coil to the RX load after the regulation) and the magnitude of power delivery to the load are limited by the size of the RX coil [12,13]. Furthermore, in the case shown in Fig. 5.1 where multiple implants receive power from a single external power transmitter, it is infeasible to control the received power level at each implant individually by adjusting the shared transmitted power. As such, the spread in distances between the external transmitter and the implants (and thereby, the range of coupling coefficients) becomes an efficiency limiting factor. Therefore, particularly for multiple mm-sized WPT implants with on-chip coils, it is important to maximize both the WPT system efficiency and magnitude of delivered power.

It can be shown that, to a first order, maximum efficiency and maximum power in a WPT system, when operating at low coupling coefficients, k, are achieved at the same optimal load of the secondary LC tank, R_{AC_opt} [17]. Specifically,

$$R_{AC\_opt} \approx \frac{L_S}{C_S R_S} \approx Q_{coil}^2 R_S, \tag{5.1}$$

where L_S, C_S, and R_S are inductance and capacitance of the secondary LC tank, and the parasitic series resistance of the RX coil as shown in Fig. 5.2; Q_{coil} is $\omega_o L_S / R_S$ where $\omega_o \approx 1/\sqrt{L_S C_S}$ at the parallel tuned LC tank.

By a series-to-parallel impedance transformation [18,19], the output impedance of the parallel tuned LC tank is the same R_{AC_opt}. Hence, both the efficiency and the amount of power are maximized through *impedance matching* between the receiver LC tank and the subsequent IR^3 circuitry. Several recent designs control matching capacitance or inductance in the secondary LC tank to adjust R_{AC_opt} according to Eq. (5.1), either to improve power transfer efficiency [7,20] or to increase the amount of delivered power [17]. These improvements come at the expense of some additional implant area and power loss due to the addition of series switches connecting to several separate capacitors in parallel or the inductor in series.

The IR^3 equivalent input resistance, R_{AC} shown in Fig. 5.2, can be found by relating the AC power consumption at its input $P_{in} = v_{rf\_in\_peak}^2 / 2R_{AC}$, where $v_{rf\_in\_peak}$ is the peak voltage in the secondary LC tank, to the output DC power $P_L = V_{DD}^2 / R_L$ delivered to a load with resistance R_L. In terms of the power conversion efficiency, PCE $= P_L/P_{in}$, and voltage conversion efficiency, VCE $= V_{DD} / v_{rf\_in\_peak}$ of the combined rectifier and regulator, this becomes

$$R_{AC} = \frac{PCE}{2VCE^2} R_L. \tag{5.2}$$

Eq. (5.2), together with (5.1), underscores the importance of maintaining high VCE in maximizing overall WSE through impedance matching. Indeed, typical on-chip inductors with quality factor $Q_{coil} = 12$ and series resistance $R_S = 3.5\ \Omega$ yield an optimal LC tank load R_{AC_opt} of roughly 500 Ω. In turn, a typical ECoG application [1,2] incurs a load $R_L = 2$ kΩ at PCE = 0.5. Under these conditions, the equivalent input resistance R_{AC} (5.2) is perfectly matched to this optimal load when VCE = 1. Conversely, when VCE = 0.2, R_{AC} becomes 12.5 kΩ, substantially off from the optimal load R_{AC_opt}. Hence, in general, mm-sized WPT receivers with less than 1 mW DC load require not only high PCE, but also high VCE.

5.2.2 Limitations of conventional cascaded rectification and regulation

Achieving both high PCE and high VCE is challenging with conventional WPT receivers which employ a cascaded two-step conversion approach: RF-DC rectification followed by DC regulation. For rectification, the simplest design is a passive rectifier built with diode-connected MOS transistors, although at low PCE and VCE due to an inherent V_{TH} voltage drop across the diodes. Alternatively, active rectifiers with high-speed comparators for active low-voltage diodes yield improved VCE, as well as PCE [21]. Both the passive and active rectifiers operate, essentially, by tracking the envelope of the RF input and pulling up the output voltage whenever it is lower. As such, the rectified output tends to the peak voltage of the RF input *regardless of the required output voltage*. Hence, variations in the received RF input amplitude due to changes in coupling conditions (and thereby, coupling coefficient) between the TX and RX coils during movement require regulation of the rectified voltage for stable operation of the implant circuits. A conventional WPT receiver accomplishes this through a separate additional regulation stage. For this purpose, low-drop-out (LDO) regulators have been widely adopted [1,22,8,9,7].

The separation between rectification and regulation functions incurs several inefficiencies as illustrated in Fig. 5.2(A). To support this two-step conversion, two large supply-decoupling capacitors are required before and after regulation to reduce voltage ripple and improve regulation feedback stability. Full on-chip integration of two large capacitors in a mm-sized implant is prohibitive not only because of the large silicon area required, but also because of eddy currents induced by the RF magnetic field in large solid metal planes which substantially reduce WSE. High-speed LDO regulators permit lowering the size of the two decoupling capacitors [22], though this is achieved by increasing bandwidth of the LDO and therefore quiescent power consumption.

In addition, the cascaded rectification-regulation two-step conversion leads to multiplicative losses in conversion efficiencies: $\eta_{REC\_REG} = \eta_{REC} \times \eta_{REG}$, both for VCE and PCE. This is particularly problematic with LDO linear regulators for which both the VCE_{REG} and PCE_{REG} are limited by the ratio of regulated to rectified voltage:

$$\text{PCE}_{\text{REG}} \leq \text{VCE}_{\text{REG}} = \frac{V_{\text{DD}}}{V_{\text{REC}}}, \tag{5.3}$$

which can be arbitrarily low depending on RF input conditions. In the case of a single implant, regulator efficiencies can be improved by dynamically adjusting the transmitted RF power at the external antenna to control V_{REC} slightly greater than V_{DD} [10,11]. However, in the case of multiple implants as depicted in Fig. 5.1, the minimum level of transmitted RF power is set by the weakest link (generally, the farthest implant from the external transmitting antenna). As a result, a uniformly high VCE and PCE cannot be guaranteed across all implants when using cascaded conversion with LDO linear regulators. Furthermore, it is important to note that the drop-out voltage ($V_{\text{REC}} - V_{\text{DD}}$) is generally greater than a few hundred mV and increases its portion as the supply voltage is scaled down [23]. As such, a LDO regulator is the most static power consuming block in several mm-sized implantable designs [8,24].

5.3. Integrated resonant rectification and regulation

To address the disadvantages of cascaded two-stage conversion, we proposed a fully integrated solution to combine rectification and regulation by directly coupling the on-chip resonant tank to the on-chip load via hybrid pulse width and frequency modulation of the conductive path between the tank and the load.

5.3.1 Benefits of integration

Combination of rectification and regulation functions within a single stage offers greater integration density as well as high efficiencies [16,25–28]. For further improvement in size and efficiencies, the RX coil is also integrated on-chip such that the integrated resonant regulating rectifier (IR^3) presented here directly converts the induced RF power at the integrated coil to the regulated DC supply driving on-chip loads without cascaded losses in overall VCE and PCE, as illustrated in Fig. 5.2(B). Hence, the IR^3 is capable of attaining high PCE and VCE not limited by the two-stage inefficiencies in (5.3). In particular, owing to improved overall VCE (greater than 90%, Fig. 5.13) by combining two functions into a single stage, the equivalent input resistance of the IR^3, R_{AC}, is close to the optimal LC tank load, $R_{\text{AC\_opt}}$, leading to improved overall WSE. While not implemented here, real-time $R_{\text{AC\_opt}}$ tracking functionality can be included with IR^3 for further improvement in overall WSE.

In addition to removing cascaded losses in efficiencies, full monolithic integration also reduces parasitic capacitance and inductance in the LC tank interfacing to the IR^3. To support full integration in an mm-sized implant, decoupling capacitance is reduced to 1 nF, which is adequate for regulation within 5 mV ripple (Fig. 5.11).

5.3.2 Hybrid pulse modulation

The IR^3 performs simultaneous rectification and regulation by controlling the amount of power transferred from the RF input to the regulated DC output, V_{DD}, through a hy-

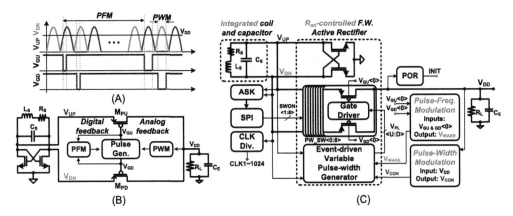

Figure 5.3 Hybrid pulse modulation (HPM) with combined pulse-width modulation (PWM) and pulse-frequency modulation (PFM) for integrated resonant rectification and regulation. (A) Conceptual operation and timing diagram. (B) Analog PWM and digital PFM feedback loops. (C) Block-level IR^3 circuit diagram.

brid pulse modulation (HPM) scheme that combines pulse-width modulation (PWM) and pulse-frequency modulation (PFM), illustrated in Fig. 5.3(A). Each pulse activates a conductive path between the resonant tank and the load, accomplishing both rectification and regulation in one step. For rectification, the pulse generator in Fig. 5.3(B) activates the up-rectifying power PMOS transistor, M_{PU}, by lowering V_{GU} when its RF input, V_{UP}, exceeds V_{DD}. Conversely, the down-rectifying power PMOS transistor, M_{PD}, is activated by lowering V_{GD} when its RF input, V_{DN}, exceeds V_{DD} in the opposite RF phase for full-wave rectification. For regulation, the width of the activation pulse is controlled by PWM through analog feedback [29], and its pulse frequency is controlled by PFM through digital feedback.

The PWM module regulates V_{DD} simultaneously with rectification, by adjusting the pulse-width based on comparison of V_{DD} with a predefined reference voltage. Owing to large loop gain at DC in purely analog feedback (Fig. 5.7 and Eq. (5.5)), the PWM mode offers accurate regulation of V_{DD}. However, due to a dominant pole compensation for stability of the analog feedback, the response time of PWM analog feedback is relatively slow, in the range of several hundred microseconds. In addition, due to timing constraints in rectification of the RF input, pulse widths themselves are very short, in the range of a few nanoseconds at tank resonant frequencies above 100 MHz.

Hence, considering typical rise and fall times in V_{GU} and V_{GD} in the upper hundreds of picoseconds range, the dynamic range of pulse-width control is very limited, and PWM alone is insufficient for accurate regulation across varying input and output power conditions. Furthermore, narrow pulse widths prohibit the IR^3 from achieving high PCE at light load conditions.

To address these challenges and provide rapid digital feedback covering wide dynamic range, another PFM regulation loop is included in addition to and in tandem with PWM. The PFM module increases the pulse frequency of V_{GU} and V_{GD} when either pulse width being regulated by PWM reaches an upper threshold; conversely, the pulse frequency is decreased when either pulse width reaches a lower threshold. As such, PFM together with PWM provides broad-range and rapid regulation and improved power conversion efficiency at light load conditions [30,31].

Fig. 5.3(C) shows the block-level circuit diagram of the IR^3. The analog output V_{CON} of the PWM module and the digital output V_{WAKE} of the PFM modules control the width and frequency of the pulsed waveforms $V_{PL<U>}$ and $V_{PL<D>}$ driving the gates of up to nine parallel-connected power PMOS transistors. The LC tank recovered clock signal is binary divided to provide various clocks (CLK1−1024) for the entire system including the PFM module and other on-chip loads. A power-on-reset generates an INIT signal forcing the IR^3 into passive mode with diode-connected PMOS transistors and initializing digital state variables in the PFM module during start-up [21]. Circuit implementation and operation of various blocks are detailed in the following section.

5.4. Circuit implementation

5.4.1 Event-driven variable pulse-width generator

The schematic of the event-driven variable pulse width generator is shown in Fig. 5.4 for half of a cycle (i.e., the positive phase of full-wave rectification). This block produces the pulse waveform $V_{PL<U>}$ from the analog PWM control V_{CON} and event-driven digital PFM control V_{WAKE} shown in Fig. 5.3(C). It comprises a fast comparator defining the turn-on time, t_{ON}, when V_{UP} exceeds V_{DD}, followed by a triggered (i.e., zero idle power) variable delay element defining the pulse width, t_P, and synthesizes the overall pulse through a NAND gate. Voltage control V_{CON} over the delay variable t_P establishes PWM control over a 0.4–2.5 ns range. For PFM control and to reduce power

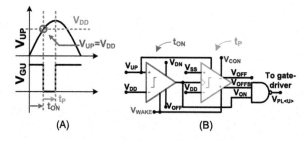

Figure 5.4 The event-driven variable pulse-width generator driven by V_{CON} PWM and V_{WAKE} PFM controls. (A) Timing diagram. (B) Simplified circuit diagram. Only the upper half (V_{UP} controlling V_{GU}) of the circuit is shown; the other half (V_{DN} controlling V_{GD}) is identical.

consumption, the entire pulse generator circuit is gated by the V_{WAKE} digital signal, which powers up the comparator and delay element only when actively pulsed. The PFM power gating is especially effective at light loads to alleviate active power by lowering the operating frequency of the pulse generator and gate driver.

Low-power design of the t_{ON} comparator is critical in the overall energy efficiency of the WPT receiver, as it operates at up to 144 MHz. Prior-art high-performance comparators consume on the order of 100–800 μW at 13.56 MHz [21]; a similar approach at 144 MHz would lead to prohibitively high comparator power in the 1–10 mW range, drastically limiting the power conversion efficiency PCE [32]

$$PCE = \frac{P_L}{P_L + P_{comp} + P_{other}}, \tag{5.4}$$

where P_L, P_{comp}, and P_{other} are the portions of power consumed by the DC load, the comparator, and other blocks, respectively. For example, a typical DC load of $P_L = 10$ μW with conventional comparator power $P_{comp} = 1$ mW would limit PCE below 1% even if the power consumed by all other blocks were negligible.

For a workable alternative offering fast decision at low-power consumption, a bias-point-assisted dynamic comparator is proposed in Fig. 5.5(A). The proposed dynamic t_{ON} comparator is active only when V_{WAKE} goes high to eliminate static power consumption. Since V_{WAKE} rises all the way to V_{DD}, dynamic t_{ON} comparison is accomplished by detecting the time at which V_{UP} exceeds V_{WAKE}. As V_{UP} reaches its peak, V_{ON_PRE} also approaches its optimal bias point for fast and accurate comparison, with V_{PX} near the logic-threshold of the following inverter chain (as illustrated in Fig. 5.5(C)). To compensate for gate driver delay (around 200 ps in simulation), the logic threshold V_{LTH} of the first inverter is adjusted for a faster decision. Further dynamic enhancement in comparator speed is accomplished with a current-reuse active g_m cell composed of M_{1N} and M_{2N}, in addition to the PMOS pair, M_{1P} and M_{2P}, to dynamically increase the output current to $2i_{acN} + i_{acP}$ for fast voltage slew in V_{ON_PRE}, while retaining low DC current consumption. In contrast, current state-of-the-art comparators limit the available current for voltage slew to i_{acP}. As such, the simulated average power consumption of the comparator reduces to 0.15–1.5 μW from a 0.8 V supply across all operating frequencies up to 144 MHz.

The variable delay circuit generates a pulse of width t_P as controlled by V_{CON}, when triggered by V_{ON}. Its schematic is shown in Fig. 5.5(B). Here, V_{CON} controls the transconductance of a dynamic latch triggered by V_{ON} to provide precise short delays (0.4–2.5 ns). Simulated averaged power consumption of the delay element for 1.5 ns t_P is 0.056–1.32 μW from a 0.8 V supply, across all operating pulse frequencies from 4.5 to 144 MHz.

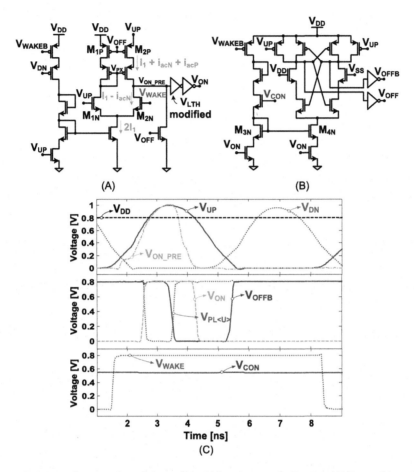

Figure 5.5 Circuit implementation of the pulse width generator in Fig. 5.4. (A) Event-driven dynamically biased t_{ON} comparator, (B) triggered variable delay t_P element, and (C) simulated time domain waveforms.

5.4.2 Pulse-frequency modulation (PFM) control module

The block diagram of the PFM module is illustrated in Fig. 5.6 along with operational timing diagrams. Under light load conditions, pulse frequency is decreased to counteract critically narrow pulse widths that would otherwise limit the resolution in PWM mode, thereby reducing the overall switching power losses (left side of Fig. 5.6(A)) Conversely, under heavy load conditions, pulse frequency is increased to avoid reverse currents from load to the tank at critically wide pulse widths (right side of Fig. 5.6(A)). The PFM module considers either critical action by monitoring pulse-width on a periodic 140 kHz V_{SAFE} schedule. Correspondingly, at the first rising edge of the active-low pulse V_{GU} when V_{SAFE} is active as shown in Fig. 5.6(A), the latch enabler in Fig. 5.6(B) triggers two latched comparators to detect two types of threshold events on the pulse-

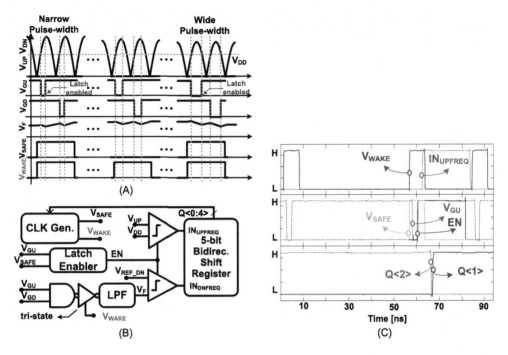

Figure 5.6 Pulse frequency modulator (PFM) module supplying the periodic V_{WAKE} signal gating the pulse generator for rapid digital feedback. (A) Pulse-frequency is decreased at a critically narrow pulse-width reaching a minimum threshold, while pulse-frequency is increased at a critically wide pulse-width, where V_{UP} (or V_{DN}) reaches below V_{DD}. (B) PFM module block diagram. (C) Simulated time domain waveforms of the PFM module when pulse-frequency is increased.

width. The PFM module considers either critical action on a periodic V_{SAFE} schedule, sufficiently slower than the PWM analog feedback response time, by monitoring pulse width once every V_{SAFE} cycle (on the rising edge of EN in Fig. 5.6(C)). Critically wide pulse widths are detected by comparing V_{UP} against V_{DD}, thereby avoiding reverse current by maintaining $V_{UP} \geq V_{DD}$ throughout pulse activation. The two comparators are latched on the EN signal to reduce power and to control the updates in pulse frequency by shifting a one-hot representation, Q<0:4>, in a 5-bit bidirectional shift register. The pulse frequency of V_{WAKE} in the clock generator is decided based on 5-bit outputs Q<0:4> of the shift register. The PFM control module shown in Fig. 5.6(B) in turn controls the operating frequency of the IR^3. The simulated time domain waveforms for a critically narrow pulse width case are shown in Fig. 5.6(C).

5.4.3 Pulse-width modulation (PWM) control module

The PWM module shown in Fig. 5.7 provides a feedback signal, V_{CON}, to the rectifier core, which contains an event-driven variable pulse generator and power transistors.

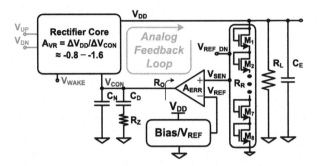

Figure 5.7 Pulse-width modulation (PWM) module for precise analog feedback.

A voltage divider, comprising a double chain of $4 + 4$ nMOS transistors, used to construct a centered sensing voltage as half of the supply voltage $V_{SEN} = V_{DD}/2$, at the expense of 6 dB loop gain loss. The quiescent current of the divider is around 40 nA, resulting in $R_R \approx 20$ MΩ at $V_{DD} = 0.8$ V. Negative feedback around the error amplifier, A_{ERR}, comparing V_{SEN} with a locally generated reference, V_{REF}, regulates the rectified supply V_{DD} near $2\,V_{REF}$. The current consumption of the reference generator block for V_{REF} is 600 nA when V_{DD} is 0.8 V.

Large loop gain plays a significant role in reducing the regulation error. As such, a two-stacked folded-cascode amplifier offering 68 dB open-loop gain at DC consuming 240 nA at V_{DD} of 0.8 V is chosen for the error amplifier. Stability of the analog feedback in supply regulation requires both dominant-pole and zero compensation in the loop. The dominant pole is set by an inserted capacitance, C_D, loading the error amplifier along with its output impedance, R_O. Asserting the dominant pole independent of load conditions requires that $R_0 C_D \gg R_{L,MAX} C_E$, where $R_{L,MAX}$ is the minimum load from the supply and C_E is the supply decoupling capacitance. For sufficient phase margin, a zero is inserted at $f_Z = 1/2\pi R_Z C_D$ by adding a resistor R_Z in series with C_D. To mitigate the resulting increase in high-frequency noise, an additional small capacitor, C_N, is inserted in parallel with C_D and R_Z, contributing a non-dominant pole at $f_N = 1/2\pi R_Z C_N$. The small-signal voltage gain of the rectifier core, A_{VR}, ranges between -0.8 and -1.6 depending on input and load conditions through PFM in the V_{WAKE} waveform. First-order analysis gives the loop gain of

$$A_L = \frac{1/2 \cdot A_{ERR} \cdot A_{VR} \cdot (1 + S R_Z C_D)}{(1 + S R_O C_D)(1 + S R_Z C_N)(1 + S R_L C_E)}. \tag{5.5}$$

Accommodating a broad range of load resistances, R_L, the load pole $f_L = 1/2\pi R_L C_E$ varies from 4 to 160 kHz as illustrated in Fig. 5.8. Owing to the dominant pole and zero compensation, the phase margin is guaranteed greater than 48° for loads heavier than 40 kΩ, ensuring stability of the analog feedback over the operating range.

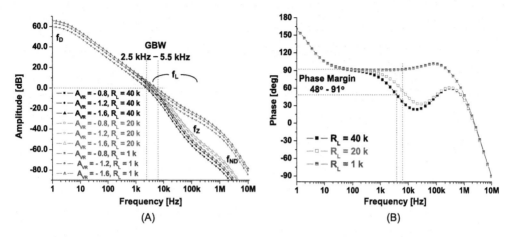

Figure 5.8 Loop gain of regulator feedback with the PWM of Fig. 5.7. Simulated amplitude (A) and phase (B) are obtained with $A_{ERR} = G_m R_O = 68.2$ dB, $C_D = 22$ pF, $R_Z = 88.6$ kΩ, $C_N = 0.6$ pF, $C_E = 1$ nF, and $R_L = 1$–40 kΩ.

5.5. Hybrid pulse modulation under RF input variation

To prove efficacy of the proposed HPM, the IR^3 is simulated with a time-varying RF input voltage as shown in Fig. 5.9. Here, a 144 MHz RF input is amplitude modulated with a 0.1 AM modulation index at 0.4 kHz, resulting in an amplitude variation from 0.98 to 1.1 V at the half-waved rectified voltage, V_{UP} and V_{DN}. After V_{DD} initially develops owing to a passive diode-connected rectification mode, the HPM starts rectification and regulation simultaneously on V_{DD} by changing, first, V_{CON} for the pulse-width and, later, Q<1:4> for the pulse-frequency as depicted in Fig. 5.9(A). This can be more clearly seen in Fig. 5.9(B)–(C). At decreasing RF input voltages, the PWM control circuit accordingly widens the pulse width via analog feedback at the fixed Q<1> mode. At its maximum pulse-width shown in Fig. 5.9(B), the PFM increases the pulse-frequency to the Q<0> mode. After reaching its minimum, the RF input voltage begins to increase again. As such, the pulse-width is getting narrowed and the pulse-frequency is back to Q<1> mode for regulation under increasing RF input. When critically narrowed pulse-width is detected by the PFM module, the pulse-frequency is increased to the Q<2> mode to avoid degradation in PCE and failure of regulation.

5.6. Measurement results

A microphotograph of an electrocortical neural interface chip [2] internally powered by the presented IR^3 is shown in Fig. 5.10(A). The active area of the IR^3 part of the chip, presented here, is 0.078 mm^2 in 180 nm 1P4M SOI. An on-chip RX coil is

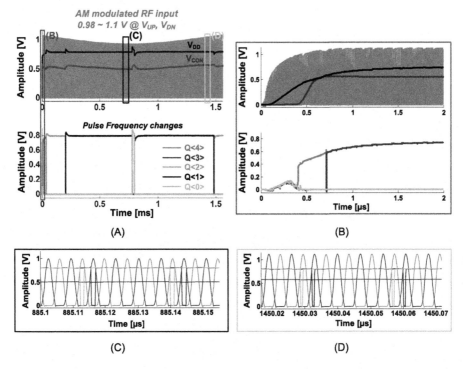

Figure 5.9 Integrated rectification and regulation with HPM under amplitude modulated (AM) RF input. (A) Pulse-frequency is dynamically adjusted based on the pulse-width by changing Q<0:4>. Colored outlined insets show (purple, B) initial power build-up waveforms, and (brown, C) pulse-frequency and pulse-width waveform detail at low RF input voltage, and (cyan, D) at high RF input voltage.

implemented at the chip edge so as to maximize its area with two turns of top metal for 23.7 nH of inductance. The top metal width of the coil is 100 μm to ensure a small series resistance, R_S. As such, HFSS electromagnetic simulation shows Q_{coil} of the RX coil is 12 at 144 MHz. As shown in Fig. 5.10(B), the rectifier core in the IR^3 is placed physically adjacent to a matching resonant capacitor to reduce parasitics. To mitigate high-frequency switching noise, sensitive analog blocks in the PWM module are placed far from the rectifier core.

For isolated characterization of RF input regulation, load regulation, and conversion efficiencies of the IR^3 circuit independent of the integrated RX coil, a PCB RF transformer (TC1-1G2+, Mini-Circuits) is initially used to bypass the on–chip LC tank and directly supply V_{UP} and V_{DN} as differential RF sinusoidal signals from a vector signal generator (Keysight N5181A).

Fig. 5.11 shows measurements of the IR^3 regulating V_{DD} at 0.8 V within 5.2 mVpp ripple at various RF input levels representative of typical link distance variations, with correspondingly varying frequency modulations. As expected, the rate of rectification

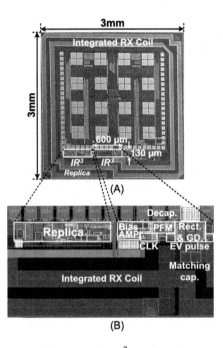

Figure 5.10 (A) Microphotograph of the presented IR^3 integrated as part of a mm-sized electrocortical neural interface chip [2] with recording, stimulation and communication circuits comprising the IR^3 load, and (B) layout detail of the IR^3 part.

(i.e., PFM frequency) is maximum (i.e., 144 MHz) at lowest RF input voltage amplitude, as the rectifier goes active every RF cycle (maximum PFM) to extract as much power from the input as possible (see Fig. 5.11(A)). Conversely, the rectifier skips several cycles (lower PFM) at high RF input levels (see Fig. 5.11(B)–(C)), as a sufficient amount of energy can be extracted from the input over fewer RF cycles. To characterize ripple, FFT spectra of the regulated voltage V_{DD}, shown at the bottom of Fig. 5.11 in each case (A) through (C), reveal a peaking tone at the pulse-frequency. Note that 144 MHz RF energy coupling through parasitics on the PCB also partially couples to V_{DD} such that Fig. 5.11(A) shows slightly higher ripple voltage than Fig. 5.11(B) despite higher pulse frequency. Altogether, the measurements demonstrate the effectiveness of the proposed hybrid pulse modulation technique under various RF input voltages.

The presented IR^3 powers a neural interface system having very different static power consuming levels according to two main operational modes, stimulation vs. recording [2]. To evaluate the capability of the IR^3 accommodating a large dynamic range in static power levels, V_{DD} is measured while changing I_{LOAD} from 8 μA to 80 μA. Owing to the proposed HPM in the IR^3 regulation functionality, a ten-fold change in I_{LOAD} incurs less than 15 mV static variation in V_{DD} as shown in Fig. 5.12(A). This static variation could be further improved by increasing loop gain with a higher gain error

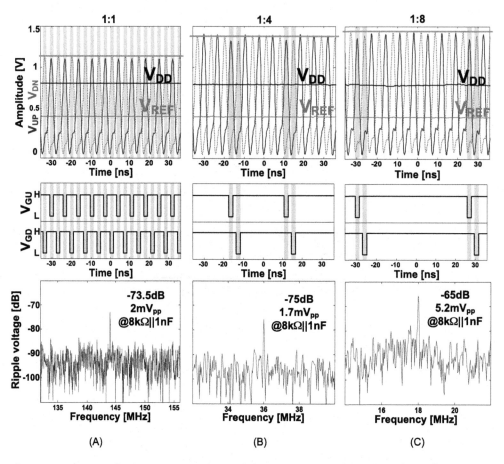

Figure 5.11 Measured RF inputs, V_{UP} and V_{DN}, and DC-regulated output V_{DD} under 8 kΩ ∥ 1 nF load operating in (A) Q<0>(144 MHz) mode, (B) Q<1>(144 MHz/4) mode, and (C) Q<2>(144 MHz/8) mode. Arrows indicate active cycles and show RF input loading at the regulation pulse frequency. Spectra for voltage ripple in the regulated output V_{DD}, peaking at the pulse frequency, are shown below for each.

amplifier. PFM effectiveness in transient regulation response under rapid load transitions is shown in Fig. 5.12(B)–(C).

Since the IR^3 performs rectification and regulation simultaneously, voltage and power conversion efficiencies from *both rectification and regulation* are reported as shown in Fig. 5.13(A)–(B). Compared to advanced active rectifier designs [32,33] achieving around 80% of VCE from only rectification, greater than 80% of VCE is observed across the range of load conditions at various target supply voltage V_{DD} levels, controlled by an externally supplied, bypassed reference V_{REF}. At $V_{DD} = 1$ V, the lowest measured VCE is 92%. As shown in Fig. 5.13(B), PCE is higher than 30% even under

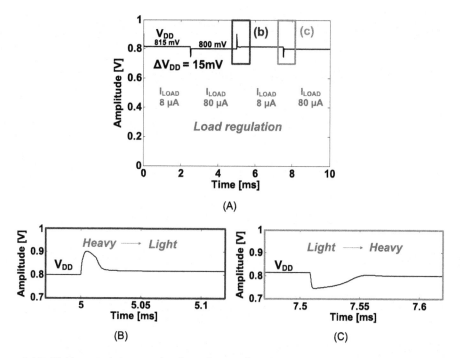

Figure 5.12 (A) Measured dynamic load regulation of V_{DD} with I_{LOAD} alternating between 8 μA and 80 μA with a 1 nF decoupling capacitor at V_{DD} node. PFM improves the regulation speed by changing the pulse-frequency at heavy-to-light load transition (B) and light-to-heavy load transition (C).

8 μW load. As such, the total power consumption of the IR^3, including power losses from the power transistors, is less than 20 μW at 144 MHz. The low power consumption is achieved mainly because of the proposed bias point assisted dynamic t_{ON} comparator. PCE improves at increasing load currents since the portion of the power consumed by control blocks of the IR^3 are independent of load changes. At maximum load, the integral PCE of the IR^3, combining rectification and regulation, reaches 60%. In contrast, power converters in state-of-the-art mm–sized implants [7,8] achieve around 60% of PCE for rectification, in cascade with 50% to 80% of PCE for regulation, for a net 30% to 50% combined PCE.

The wireless inductive link efficiency and overall WSE with the RX integrated on–chip coil connected to the IR^3 are characterized using a 2.5 cm × 2.5 cm, single-turn TX coil integrated with a passive matching network on a printed circuit board (PCB) as depicted in Fig. 5.14(A). Quality factor of the TX coil is greater than 70 at the resonance frequency. The TX coil inductively couples to the on–chip two-turn (2.4 mm × 2.4 mm inner-loop, 2.9 mm × 2.9 mm outer-loop) RX coil over 1 cm distance as shown in Fig. 5.14(B). In this setup, S-parameters between the TX coil and the RX integrated on–chip coil are measured with a network analyzer (Keysight,

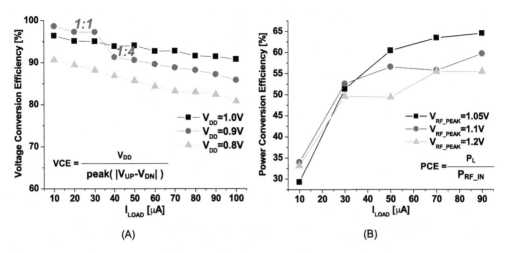

Figure 5.13 Measured voltage conversion efficiency (A) and simulated power conversion efficiency (B) of the IR^3 under varying external load current, target regulation voltage, and RF input conditions.

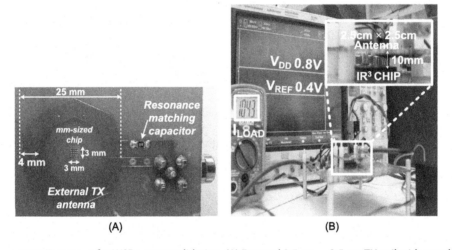

Figure 5.14 Test setup for WSE system validation. (A) External 2.5 cm × 2.5 cm TX coil with matching network. The mm-sized implant chip with the presented IR^3 shown in Fig. 5.10(A) is superimposed in the center for size comparison. (B) Wireless power transfer at 1 cm distance between TX and on-chip RX coils.

E5080A), and converted to Z-parameters. Parasitics from the matching network, PCB traces, SMA connectors, and cables are measured and de-embedded [6,34] to obtain the accurate link efficiency showing the net power gain from the transmitted power at the TX coil (P_{TX}) to the received power at a 4.3 kΩ AC load R_{AC} to the RX coil. According to Eq. (5.2), 4.3 kΩ of AC load R_{AC} is equivalent to 8.9 kΩ of R_L with 0.82 of VCE and 0.65 of PCE (Fig. 5.13), rendering a 90 µW load to V_{DD}. With this load,

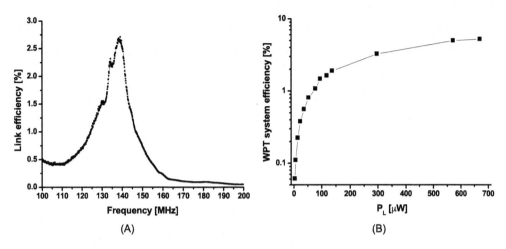

Figure 5.15 Measured WPT power efficiencies with external 2.5 cm × 2.5 cm TX coil. (A) Link efficiency between the external TX coil and the RX 2-turn on-chip coil with AC resistance $R_{AC} = 4.3$ kΩ. (B) WPT system efficiency, WSE $= P_L/P_{TX}$ at 1 cm distance between TX and on-chip RX coils. The IR^3 maintains a constant 800 mV V_{DD} for loads up to 700 μW.

the peak link efficiency across various RF frequencies is 2.7% at the RX coil resonance frequency as shown in Fig. 5.15(A). Measured overall WSE (P_L/P_{TX}) under various load conditions is shown in Fig. 5.15(B). The IR^3 maintains a constant 800 mV V_{DD} for loads up to 700 μW. Greater than 2% overall WSE from TX coil to the regulated DC is demonstrated with a 160 μW load over air while, at the same input power to the TX coil, 8 dB additional power loss is measured in a more realistic biomedical setting implemented by inserting 1 cm of porcine fatty abdominal tissue.

Table 5.1 summarizes key performance measures of the IR^3 in comparison to state-of-the-art designs for mm-sized implants. Among these, only IR^3 utilizes a regulating rectifier to improve PCE and, thus, WSE. The tabulated figure-of-merit (FoM) is based on one proposed by [12], considering that for mm-sized RX coils (in the mid-field regime) the TX-RX link efficiency is inversely proportional to the cube of the inter-coil distance [13] and the cube of square root of RX coil area [12]. In the comparison of Table 5.1, we extend this FoM from TX-RX link efficiency to overall WSE also accounting for the implant PCE of the form

$$\text{WSE FoM} = \frac{\text{WSE} \times D^3}{A^{1.5}} \tag{5.6}$$

where D is the distance between TX and RX coils, and A is the outer-loop area of the RX coil.

Table 5.1 Performance comparison.

	[1] Muller	[7] O'Driscoll	[8,9] Mark	[12,22] Zargham	This work
RX coil	Off-chip/ 32 nH	Off-chip/ N/R	Off-chip/ 5.73 nH	On-chip/ 130 nH	On-chip/ 23.7 nH
Area of RX coil [mm^2]	42.25	4	1	4.36	8.64
Res. Freq. [MHz]	300	915	535	160	144
Regulator	Separated LDO	Separated LDO	Separated LDO	Separated LDO	Regulating rectifier
Regulation[*] [%]	N/R	N/R	N/R	N/R	1.87
Decoupling Cap. [nF]	4	N/R	1.39	0.02	1
Process	65 nm 1P7M CMOS	0.13 μm CMOS	65 nm CMOS	0.13 μm 1P8M CMOS	0.18 μm 1P4M CMOS SOI
Overall WSE (TX to V^{DD})[%]	1.19[a]	0.048 (−33.2 dB)	0.02 (−37 dB)	0.62[b]	2.04[c]
Distance [mm]	12.5	15	13	10	10
WSE FOM[**]	8.46	20.25	43.94	68.1	80.3

[a] Estimated from provided data:
   transmit PW: 13 mW, received PW: 0.225 mW (simulated link gain, −16.5 dB) and P$_L$: 0.160 mW.
[b] Estimated from provided data:
   provided efficiency from TX to the output of rectifier: 0.9%, calculated efficiency of LDO: 68% (V$_{DD}$: 3.1 V, V$_{REC2}$: 4.5 V).
[c] Measured TX power: 7.87 mW and P$_L$: 0.160 mW.
[*] Defined by [25], Regulation = Static ΔV$_{DD}$ / nominal V$_{DD}$.
[**] Modified from [12], WSE FOM = $\dfrac{\text{Overall WSE} \times \text{Distance}^3}{\text{Area of RXcoil}^{1.5}}$.

5.7. Conclusions

This chapter demonstrated the first fully integrated mm-sized WPT receiver for micropower mm-sized biomedical implants operating without off-chip components. Higher resonant frequencies, such as 144 MHz demonstrated here, support full integration of the RX coil with reduced size matching capacitor, and offer higher link efficiency. Crucially, the integration of rectification and regulation yields substantially higher PCE and VCE along with savings in silicon area by avoiding the need for decoupling capacitors in mm-sized implants. The unique combination of RF inductive resonant power transfer, rectification, and hybrid PWM-PFM regulation of the IR^3 offers superior voltage and power conversion efficiency alleviating severe powering conditions of deep mm-size biomedical implants. As suggested in Fig. 5.1, we envision tiled arrays of mm-sized implants distributed across the surface of cortex with fully integrated IR^3s for power reception and with data transceivers for data communication, performing simultaneous parallel power delivery and data telemetry using a single external loop antenna driven by a beamforming transceiver IC [35,36].

Acknowledgment

The authors thank Shadi Dayeh, Vikash Gilja, Eric Halgren, Bruce McNaughton and other researchers on the ENIAC Memory Prostheses team for valuable interactions and collaborations.

References

[1] R. Muller, H.-P. Le, W. Li, P. Ledochowitsch, S. Gambini, T. Bjorninen, A. Koralek, J. Carmena, M. Maharbiz, E. Alon, J. Rabaey, A minimally invasive 64-channel wireless μECoG implant, IEEE Journal of Solid-State Circuits 50 (1) (Jan. 2015) 344–359.

[2] S. Ha, A. Akinin, J. Park, C. Kim, H. Wang, C. Maier, P.P. Mercier, G. Cauwenberghs, Silicon-integrated high-density electrocortical interfaces, Proceedings of the IEEE 105 (1) (Jan. 2017) 11–33.

[3] H. Lee, R.V. Bellamkonda, W. Sun, M.E. Levenston, Biomechanical analysis of silicon microelectrode-induced strain in the brain, Journal of Neural Engineering 2 (4) (Dec. 2005) 81–89.

[4] G.C. McConnell, H.D. Rees, A.I. Levey, C.-A. Gutekunst, R.E. Gross, R.V. Bellamkonda, Implanted neural electrodes cause chronic, local inflammation that is correlated with local neurodegeneration, Journal of Neural Engineering 6 (5) (Aug. 2009).

[5] L. Karumbaiah, T. Saxena, D. Carlson, K. Patil, R. Patkar, E.A. Gaupp, M. Betancur, G.B. Stanley, L. Carin, R.V. Bellamkonda, Relationship between intracortical electrode design and chronic recording function, Biomaterials 34 (33) (Nov. 2013) 8061–8074.

[6] D. Ahn, M. Ghovanloo, Optimal design of wireless power transmission links for millimeter-sized biomedical implants, IEEE Transactions on Biomedical Circuits and Systems 10 (1) (Feb. 2016) 125–137.

[7] S. O'Driscoll, A. Poon, T. Meng, A mm-sized implantable power receiver with adaptive link compensation, in: IEEE International Solid-State Circuits Conference Digest of Technical Papers (ISSCC), Feb. 2009, pp. 294–295.

[8] M. Mark, Powering mm-Size Wireless Implants for Brain-Machine Interfaces, Ph.D. dissertation, Electrical Engineering and Computer Sciences, University of California at Berkeley, Dec. 2011.

[9] M. Mark, Y. Chen, C. Sutardja, C. Tang, S. Gowda, M. Wagner, D. Werthimer, J. Rabaey, A 1 mm^3 2 Mbps 330 fJ/b transponder for implanted neural sensors, in: Symposium on VLSI Circuits (VLSI Circuits), Jun. 2011, pp. 168–169.

[10] X. Li, C.-Y. Tsui, W.-H. Ki, Wireless power transfer system using primary equalizer for coupling- and load-range extension in bio-implant applications, in: IEEE International Solid-State Circuits Conference (ISSCC), Feb. 2015, pp. 228–229.

[11] X. Li, X. Meng, C.Y. Tsui, W.H. Ki, Reconfigurable resonant regulating rectifier with primary equalization for extended coupling- and loading-range in bio-implant wireless power transfer, IEEE Transactions on Biomedical Circuits and Systems 9 (6) (Dec. 2015) 875–884.

[12] M. Zargham, P. Gulak, Fully integrated on-chip coil in 0.13 μm CMOS for wireless power transfer through biological media, IEEE Transactions on Biomedical Circuits and Systems 9 (2) (Apr. 2015) 259–271.

[13] A. Poon, S. O'Driscoll, T. Meng, Optimal frequency for wireless power transmission into dispersive tissue, IEEE Transactions on Antennas and Propagation 58 (5) (May 2010) 1739–1750.

[14] M. Zargham, P. Gulak, Maximum achievable efficiency in near-field coupled power-transfer systems, IEEE Transactions on Biomedical Circuits and Systems 6 (3) (June 2012) 228–245.

[15] J.M. Rabaey, M. Mark, D. Chen, C. Sutardja, C. Tang, S. Gowda, M. Wagner, D. Werthimer, Powering and communicating with mm-size implants, in: Design, Automation Test in Europe, Mar. 2011, pp. 1–6.

[16] C. Kim, S. Ha, J. Park, A. Akinin, P. Mercier, G. Cauwenberghs, A 144 MHz integrated resonant regulating rectifier with hybrid pulse modulation, in: Symposium on VLSI Circuits (VLSI Circuits), June 2015, pp. 284–285.

[17] P.P. Mercier, A.P. Chandrakasan, Rapid wireless capacitor charging using a multi-tapped inductively-coupled secondary coil, IEEE Transactions on Circuits and Systems I: Regular Papers 60 (9) (Sept. 2013) 2263–2272.

[18] R. Sarpeshkar, Ultra Low Power Bioelectronics, University Press, Cambridge, 2010.

[19] P.P. Mercier, A.P. Chandrakasan, Ultra-Low-Power Short-Range Radios, Springer, 2015.

[20] B. Lee, P. Yeon, M. Ghovanloo, A multicycle Q-modulation for dynamic optimization of inductive links, IEEE Transactions on Industrial Electronics 63 (8) (Aug. 2016) 5091–5100.

[21] H.-M. Lee, M. Ghovanloo, A high frequency active voltage doubler in standard CMOS using offset-controlled comparators for inductive power transmission, IEEE Transactions on Biomedical Circuits and Systems 7 (3) (June 2013) 213–224.

[22] M. Zargham, P.G. Gulak, A 0.13 μm CMOS integrated wireless power receiver for biomedical applications, in: Proceedings of the ESSCIRC (ESSCIRC), Sept. 2013, pp. 137–140.

[23] W.-C. Chen, S.-Y. Ping, T.-C. Huang, Y.-H. Lee, K.-H. Chen, C.-L. Wey, A switchable digital–analog low-dropout regulator for analog dynamic voltage scaling technique, IEEE Journal of Solid-State Circuits 49 (3) (Mar. 2014) 740–750.

[24] J. Charthad, M.J. Weber, T.C. Chang, A. Arbabian, A mm-sized implantable medical device (IMD) with ultrasonic power transfer and a hybrid bi-directional data link, IEEE Journal of Solid-State Circuits 50 (8) (Aug. 2015) 1741–1753.

[25] J.-H. Choi, S.-K. Yeo, S. Park, J.-S. Lee, G.-H. Cho, Resonant regulating rectifiers (3R) operating for 6.78 MHz resonant wireless power transfer (RWPT), IEEE Journal of Solid-State Circuits 48 (12) (Dec. 2013) 2989–3001.

[26] X. Li, C.Y. Tsui, W.H. Ki, A 13.56 MHz wireless power transfer system with reconfigurable resonant regulating rectifier and wireless power control for implantable medical devices, IEEE Journal of Solid-State Circuits 50 (4) (Apr. 2015) 978–989.

[27] L. Cheng, W.H. Ki, T.T. Wong, T.S. Yim, C.Y. Tsui, A 6.78MHz 6W wireless power receiver with a 3-level 1 × / 1/2 × / 0 × reconfigurable resonant regulating rectifier, in: IEEE International Solid-State Circuits Conference (ISSCC), Jan. 2016, pp. 376–377.

[28] H.M. Lee, C.S. Juvekar, J. Kwong, A.P. Chandrakasan, A nonvolatile flip-flop-enabled cryptographic wireless authentication tag with per-query key update and power-glitch attack countermeasures, IEEE Journal of Solid-State Circuits 52 (1) (Jan. 2017) 272–283.

[29] H.M. Lee, H. Park, M. Ghovanloo, A power-efficient wireless system with adaptive supply control for deep brain stimulation, IEEE Journal of Solid-State Circuits 48 (9) (Sept. 2013) 2203–2216.

[30] B. Sahu, G.A. Rincon-Mora, An accurate, low-voltage, CMOS switching power supply with adaptive on-time pulse-frequency modulation (PFM) control, IEEE Transactions on Circuits and Systems I: Regular Papers 54 (2) (Feb. 2007) 312–321.

[31] R.W. Erickson, D. Maksimovic, Fundamentals of Power Electronics, 2nd ed., Springer US, 2001.

[32] H.-M. Lee, M. Ghovanloo, An integrated power-efficient active rectifier with offset-controlled high speed comparators for inductively powered applications, IEEE Transactions on Circuits and Systems I: Regular Papers 58 (8) (Aug. 2011) 1749–1760.

[33] G. Bawa, M. Ghovanloo, Active high power conversion efficiency rectifier with built-in dual-mode back telemetry in standard CMOS technology, IEEE Transactions on Biomedical Circuits and Systems 2 (3) (Sept. 2008) 184–192.

[34] A. Ibrahim, M. Kiani, A figure-of-merit for design and optimization of inductive power transmission links for millimeter-sized biomedical implants, IEEE Transactions on Biomedical Circuits and Systems 10 (6) (Dec. 2016) 1100–1111.

[35] C. Kim, S. Joshi, C. Thomas, S. Ha, L. Larson, G. Cauwenberghs, A 1.3 mW 48 MHz 4-channel MIMO baseband receiver with 65 dB harmonic rejection and 48.5 dB spatial signal separation, IEEE Journal of Solid State Circuits 51 (4) (Apr. 2016) 832–844.

[36] D. Seo, H.Y. Tang, J.M. Carmena, J.M. Rabaey, E. Alon, B.E. Boser, M.M. Maharbiz, Ultrasonic beamforming system for interrogating multiple implantable sensors, in: Proceedings of the Annual International Conference of the IEEE Engineering in Medicine and Biology Society (EMBC), Aug. 2015, pp. 2673–2676.

CHAPTER 6

RF power transmission and its considerations for ECoG implants

Contents

6.1. Introduction

Miniaturization of implants to millimeter dimensions as illustrated in Fig. 6.1 opens up new possibilities to long-lasting, robust, and information-rich brain–machine interface (BMI) technologies. BMI technologies with current implantable devices offering limited operation time (months to a year) have proven their potential. For instance, recent BMI research successfully demonstrates recovery from spinal cord injuries [1]. Thus, there is a great interest in the medical community for the development of long-term and unobtrusive BMIs to enable quality of life improvements for patients suffering from debilitating conditions. Miniature implants overcome a major obstacle to the longevity of BMI technologies, as reduced micro-motions result in less astroglial scarring, and cell death [2–5]. In addition, miniature implants can be modularly deployed to extend their coverage to the entire cortical surface while maintaining high spatial resolution [6,7].

The biggest component by volume in conventional implants is typically the battery. Ideally, the total size of an implant for BMI should be mainly limited by the physical requirements of the sensor and functional electronics rather than that of the energy source. Therefore, bulky batteries are not an acceptable solution to power implants for

High-Density Integrated Electrocortical Neural Interfaces
https://doi.org/10.1016/B978-0-12-815115-0.00013-1

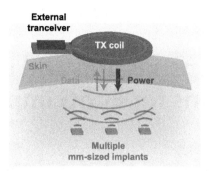

Figure 6.1 Wireless power transmission and bi-directional data communication with multiple mm-sized implants served by a single external transceiver.

BMI. Wireless power transmission (WPT) offers a viable alternative to battery power, by enabling drastic miniaturization and extended life time of the implant [8].

However, miniaturization to mm dimensions is still challenging because a plurality of up-to-date current WPT techniques still rely on bulky external power receiving (RX) coils and energy-storage decoupling capacitors. Thus, implementation of WPT with a small form factor for mm-sized implants persists as an unresolved issue. Ultimate miniaturization can be truly enabled by integrating those external bulky components into an free-floating on-chip wireless power receiver.

This chapter reviews two major WPT modalities for mm-sized implants in Sect. 6.2, and discusses main design considerations for fully integrated wireless power receivers in Sect. 6.3. In Sect. 6.5, a state-of-the-art regulating rectifier applicable to miniaturized wireless power receiver with an on-chip coil is presented with, followed by its measurement results and a performance comparison in Sect. 6.6 and conclusions in Sect. 6.7.

6.2. Wireless power transmission modalities for miniaturized implants

Two different WPT modalities have been widely investigated for mm-sized implants – WPT using either ultrasound or electromagnetic. Baseline mechanisms, advantages and disadvantages are discussed for each of these two modalities in the following section.

6.2.1 Ultrasonic wireless power transmission

For ultrasonic WPT, power transmitting (TX) and receiving (RX) parts are implemented with piezoelectric transducers for conversion between electrical and acoustic energy. On the TX side, electrical energy is converted to a pressure wave, which is then transcutaneously transmitted to the RX side (implants) through the media. The RX transducer converts part of the received acoustic energy back to electrical form [9]. Owing to relatively slow speed of propagation ($\approx$ 1540 m/s in human soft-tissue), ul-

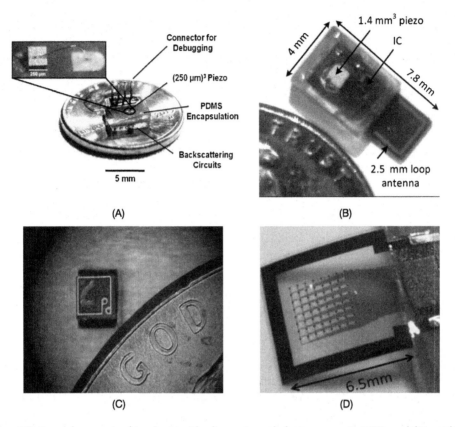

Figure 6.2 Example mm-sized implants with ultrasonic and electromagnetic WPT modalities. Ultrasonic WPTs with: (A) 250 μm^3 piezo transducer and backscattering circuits composing of several discrete components [11]; and (B) a 1.4 mm^3 piezo transducer, a 1 mm × 2 mm chip, and a 2.5 mm × 2.5 mm off-chip RF antenna for data communication [12]. Electromagnetic WPTs with (C) 1 mm × 1 mm 535 MHz RX coil connected to the same size CMOS IC beneath [13], and (D) 6.5 mm × 6.5 mm 300 MHz RX coil and 64 channel electrode array [7].

trasound offers a wavelength comparable (≈ 1.5 mm at 1 MHz of operating frequency) to the mm–sized implant dimensions. The matched wavelength and implant dimensions allow for focal power delivery resulting in efficient coupling efficiency [10]. Correspondingly, recent studies successfully demonstrated ultrasonic WPT delivery of up to 100 μW to mm–sized implants at cm-range separation between TX and RX as shown in Fig. 6.2(A) [11] and Fig. 6.2(B) [12].

Despite great advantages, applications of ultrasonic WPT to implants are limited. As the ultrasonic pressure wave travels toward the implant, it traverses several layers of tissue with substantial differences in acoustic impedances. Crossing tissue boundaries causes reflections of the pressure wave proportional in amplitude to the degree of acoustic impedance mismatch. While such reflections are highly beneficial as the key principle in ultrasonographic imaging, they deteriorate the energy coupling efficiency in WTP applications. For example, the acoustic impedance of skull is more than 4.6 times greater than that of adjacent soft tissue layers, causing a pressure reflection ratio of 0.64, and resulting in less than 2% of the incident power making it across towards a transcranial implant [14,15]. As such, ultrasonic WPT is only effective for powering implants where there is little to negligible acoustic impedance mismatch in the path from TX to the implant. The reach of ultrasonic WPT could, however, be extended, in principle, by inserting repeaters at impedance mismatch boundaries. For instance, to avoid the huge power losses due to skull reflection, a WPT hub sub-cranial interrogator can be inserted under the skull combining electromagnetic WPT to an external transceiver above, with ultrasonic WPT to mm-sized implants underneath [11]. Typical currently achievable transfer efficiencies from WPT interrogators to mm-sized implants at cm-range distances are less than 0.1% due in part to severe size limits in both piezo transducers. In addition to power delivery, data communication via ultrasound has a limited data rate due to its relatively low operating frequency ($\leq$ 10 MHz). To improve up-link (from implants to external transceiver) data rates to several Mbps, 4 GHz ultra-wideband (UWB) RF communication has been employed with an additional PCB loop antenna [12].

It is worth mentioning also that most of current state-of-the-art ultrasonic WPTs employ lead zirconate titanate (PZT) which offers high transfer efficiency owing to its superior electromechanical coupling coefficient compared to other piezo-materials. Unfortunately, PZT components included in implants pose significant health risks of long-term lead exposure inside the human body. These and other current challenges of ultrasonic WPT will likely be overcome with future research advances in materials and integrated circuit design.

6.2.2 Electromagnetic wireless power transmission

Electromagnetic (EM) WPT is currently most commonly adopted with mm-sized implants. EM WPT requires TX and RX coils to be inductively coupled through matching resonant tanks for wireless power delivery. Due to the miniature size of the implant, the tank resonance condition requires the addition of lumped capacitance to the coil inductance, requiring careful sizing considerations. The TX coil produces a time varying magnetic flux, which is shared with the RX coil where an electromotive force (EMF) is generated. Since the EMF is directly proportional to the area of RX coil by Faraday's law of induction, the size of the RX coil often dominates the size of implants for better WPT performance. The challenge with miniaturized implants is then to minimize the size of the RX coil without greatly compromising WPT performance. A common strategy is to increase the operating frequency of the implant resonant tank, which increases induction via higher time-varying rate of magnetic flux, and increases quality (Q) factors of both TX and RX coils. A higher operating frequency, however, also increases tissue absorption of the incident electromagnetic radiation, and thus reduces the maximum transmittable power to the implant under the regulations of specific absorption rate (SAR). As such, by balancing both effects, operating frequencies from 100 MHz to 1 GHz have proven optimal for EM WPT to mm-sized implants [16–18,13,19,7]. Examples of mm-sized implants utilizing EM WPT operating in this range of resonant frequencies are shown in Fig. 6.2(C)–(D).

Operating frequencies above 100 MHz open up the possibility of full integration of the RX coil directly on chip with a standard CMOS process. This integration offers significant advantages to mm-sized implants: (i) ultimate miniaturization of implants by eliminating the space previously occupied by the RX coil, (ii) no special fabrication process is needed to implement an RX coil, (iii) greatly simplified encapsulation via removing any lines and connectors between the RX coil to the electronics, and (iv) reduced parasitic capacitance and resistance between the RX coil and its fully integrated matching network (thereby eliminate additional power loss) [20–23]. Therefore, integration of RX coil on-chip is a significant step toward free-floating miniaturized implants.

6.3. Design considerations for fully integrated inductive wireless power receivers

While implementing an on-chip coil offers various advantages, it requires careful design of the wireless power receiver to optimize the WPT system efficiency (WSE), defined as the ratio of RX power delivered to the load to the TX transmitted power. WSE is an overall measure accounting for several factors in the WPT system, including two important factors on the implant side: the efficiency of the TX-RX inductive link, and the efficiency of RX RF-to-DC conversion. These two factors are described in turn in the following sections.

6.3.1 RX coil design

The maximum achievable TX-RX link efficiency, and hence WSE, depends strongly on the coupling coefficient k. To first order, the coupling coefficient k depends on coil geometry and separation distance between TX and RX coil, which are mostly given by the application [24]. Controllable design parameters in maximizing WSE are mainly the Q factors of TX and RX coils, both of which should be maximized. Optimal design of the on-chip RX coil is of paramount importance since the Q of the RX coil is limited more stringently compared to that of TX coil [16].

As a guiding principle, the on-chip RX coil is implemented with thick top-layer metal at the chip boundaries to maximize its area (for better coupling coefficient, k) and to minimize its parasitic resistance (for higher Q factor) and parasitic capacitance to substrate (for minimum losses and higher self-resonance frequency).

The number of turns of the RX coil critically affects its Q factor. At a given resonant frequency, a higher number of turns leads to a larger coil inductance and smaller matching capacitance of the LC tank. To first order, the inductance of the coil is proportional to the square of the number of turns while parasitic series resistance increases with the number of turns. As such, Q tends to increase with number of turns at certain frequency range. If large number of metal stacks are supported in the process, then series connections of two or three top-layer metal lines can be used to increase the effective number of turns within the space constraints [21]. Likewise, multiple metal layers in parallel can reduce sheet resistance of the RX coil to improve Q factor.

At large number of turns or at higher frequencies, however, Q tends to decrease rather than increase with the number of turns because the increased parasitic capacitance results in a lower self-resonance frequency. Furthermore, a many-turn coil offers larger voltage swing but limited current driving capability at the LC tank. This induces circuit design difficulties where the load implied by the implant circuits draws sparse but instantly large current out of the LC tank, such as for electrical stimulation. Supply voltage generation at the rectifier and regulator may then fail due to significant voltage drop at the LC tank. Resonance with a smaller matching capacitor is also more vulnerable to any parasitic capacitance and fabrication mismatch.

In summary, designs with larger numbers of turns for the RX coil offer greater Q for greater WSE, but at the expense of vulnerabilities to parasitics. In practice, for typical mm-sized implants operating between 100 and 300 MHz, between 2 and 4 turns are appropriate.

6.3.2 Power and signal distribution

One additional challenge in fully integrated WPT receivers arises due to the proximity of active electronics to the receiving coil. Eddy currents induced by the alternating magnetic field affect significant energy losses limiting the Q of the RF coil, and further affect the functioning of sensitive analog circuits in the implant. Eddy currents in the silicon substrate can be minimized by adopting a silicon-on-insulator (SOI) process; however, eddy currents in metal layers potentially pose a more significant problem, requiring to pay extreme attention on metal routing near the coil.

In particular, it is critically important to avoid any metal loops in the design and layout of all circuits in the implant chip. The main culprits in generating Q-reducing loops and large metal planes are often power and signal lines distributed across the entire chip such as ESD power supply lines connected to all pads or chip guard-ring metal, and congregated decoupling capacitors (decaps). These loops and metal planes induce eddy currents reducing Q factor of the RX coil by over 60% (thereby reducing WPT efficiency), while also introducing noise to sensitive circuits and a signal distribution network. To avoid large metal planes on-chip, Zargham and Gulak [20] reduced decaps to only 20 pF through inclusion of a high-performance but high-power linear regulator on the wireless

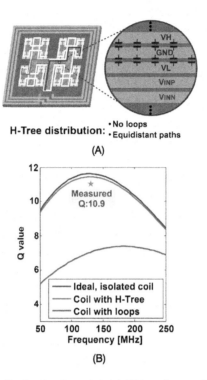

Figure 6.3 Power and signal distribution for minimal RF interference in fully integrated mm-sized WPT. (A) Fractal low-loss H-tree power and signal distribution network with energy-storage decoupling capacitors without loops and with equidistant signal paths (red and blue lines, mid gray and black in print version), and (B) simulated and measured Q factor of the RX coil [23].

power receiver. A fractal H–tree power and signal distribution network with distributed decaps shown in Fig. 6.3(A) is a practical solution to systematically remove all loops and large planes from the wireless power receiver [23]. This H-tree topology furthermore serves as a network backbone for eliminating differential mode interference in sensitive analog differential signals by ensuring equidistant signal paths. HFSS simulation and measurement in Fig. 6.3(B) show negligible loss in Q compared to an ideal, isolated coil, such that most of the incident RF energy directly couples to the RX coil rather than the metal traces cohabiting the chip. Hence, with H-tree power and signal distribution on a wireless power receiver, WSE is maximized while also decoupling RF interference.

6.3.3 Impedance matching

In addition to the H-tree power and signal distribution, impedance matching between the LC tank and subsequent electronics (typically a rectifier and a regulator) on a wireless power receiver plays a significant role in maximizing wireless power delivery and its transfer efficiency to the receiver.

The impedance of the parallel tuned LC tank becomes $Q_{coil}^2 R_{coil}$ at the resonant frequency, where Q_{coil} and R_{coil} are the quality factor and parasitic series resistance of the RX coil, respectively [24,25]. At low coupling coefficient k (most cases of mm-sized implants), both the maximum power delivery and maximum transfer efficiency are obtained at the same optimal load of the parallel tuned LC tank R_{LC_optL} given by [25]

$$R_{LC\_optL} \approx \frac{L_{coil}}{C_{matching} R_{coil}} \approx Q_{coil}^2 R_{coil}, \qquad (6.1)$$

where L_{coil} and $C_{matching}$ are inductance and capacitance in the LC tank, respectively. Thus, the optimal load impedance of the LC tank is the same to its impedance at resonance. An input impedance of a following rectifier should be matched to R_{LC_optL} to maximize both the amount of power delivery and its transfer efficiency [26].

The equivalent input resistance of the following rectifier, R_{IN}, is found by relating the RMS power consumption at its input $P_{IN} = v_{rf\_in\_peak}^2 / 2R_{IN}$, where $v_{rf\_in\_peak}$ is the peak voltage in the LC tank, to the output DC power $P_L = V_{DD}^2 / R_L$ delivered to a load R_L. In terms of the power conversion efficiency $PCE = P_L/P_{IN}$ and voltage conversion efficiency $VCE = V_{DD} / v_{rf\_in\_peak}$ of following rectification and regulation, this becomes

$$R_{IN} = \frac{PCE}{2VCE^2} R_L. \qquad (6.2)$$

Eq. (6.2), which should have the same resistance of R_{LC_optL} for impedance matching to the LC tank, highlights the importance of maintaining high VCE in maximizing overall WPT efficiency. Recent low-power sensors for BMI applications incur a load $R_L = 2$ kΩ at PCE = 0.5 [6,7]. Since the impedance of typical on–chip coil at resonance is roughly 500 Ω, VCE should be equal to 1 for R_{IN} to be perfectly matched to the optimal load. Otherwise, R_{IN} is substantially off from the optimal load. Indeed, mm-sized wireless power receivers with less than 1 mW DC load require not only high PCE, but also high VCE for maximizing the amount of power delivery and its transfer efficiency.

6.4. Rectification and regulation

Wireless link distances and alignments between the external TX coil and each implant are diverse. Therefore, the coupling coefficient, k, for each link is also individual, inducing different RF energy at the secondary LC tank of each implant. Regardless of wireless link environments, however, power supplies developed by a power management circuit should be regulated for stable operation of on-chip loads such as AFE and ADC. Hence, a power management circuit has to include functionalities of both rectification and regulation.

Cascaded rectification and regulation shown in Fig. 6.4(A) has been widely adopted since it is simple to implement and robust to wireless link variation [7,13,17,19]. How-

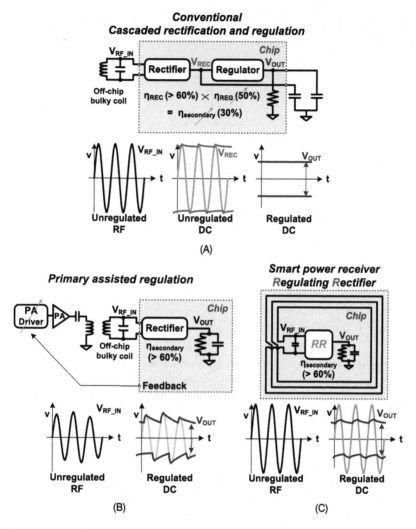

Figure 6.4 Approaches for rectification and regulation; (A) Conventional separated rectification and regulation; (B) Primary assisted regulation; (C) Regulating rectifier with no primary assistance.

ever, it should undergo cascaded loss in PCE and VCE from both rectification and regulation while requiring two huge decoupling capacitors. To remove regulators at the secondary side, primary assisted regulation method in Fig. 6.4(B) was proposed. The basis of the method is that the transmitted power at the TX coil is controlled by feedback signal from the secondary implant according to rectified output voltage V_{OUT} [27, 28]. By this approach, the primary side may be optimized to a specific wireless link environment and thereby to a specific implant. However, this implies that for multiple modular implants, multiple TX coils are required. In addition, regulation speed is possibly be limited since this feedback is composed of a gigantic loop including many delay components. For the case that multiple implants receive magnetic flux from the shared TX coil, each implant ought to be smart enough to do rectification and regulation simultaneously and locally. Hence, regulating rectifiers shown in Fig. 6.4(C) have been gradually adopted for multiple implants [22,23,29] and even for RFID authentication tags [30].

6.5. Adaptive buck-boost mode regulating rectifier

The regulating rectifier should have wide RF input-range to cover diverse wireless link distances between TX coil and implants, and offer fast regulation speed to be impervious to abrupt transitions in RF input and load, possibly from amplitude-shift-keying (ASK) RF envelope data communication and stimulation on/off, due to absence of huge off-chip decoupling capacitors. The presented adaptive buck-boost resonant regulating rectifier (B^2R^3) conducts rectification and regulation simultaneously to produce dual outputs, VH and VL. For each output, a feedback module, a buck-mode regulating rectifier (buck RR), and a boost-mode regulating rectifier (boost RR) are implemented while a mode-arbiter is shared to both outputs. The mode-arbiter with two different regulating rectifiers offers multi-mode adaptation for greater than 11 dB RF input-range. The feedback module features a dual path feedback loop for less than 0.5 μs load regulation speed.

6.5.1 Mode arbiter

Multi-mode adaptation is proposed to widen RF input range. At low RF input ($\ll 0.4$ V in amplitude), the B^2R^3 is configured to the boost mode to develop dual DC target voltages, i.e., VH and VL, from lower RF input. Conversely, at high RF input ($\gg 0.4$ V in amplitude), buck modes configuration is enabled to offer voltage-down rectification and regulation. In buck modes, three segmented power MOSFETs in the buck-mode regulating rectifier are turned on/off according to RF input voltage to alleviate power loss in the power MOSFETs, as illustrated in Fig. 6.5(A). Mode arbiter shown in Fig. 6.5(B) features automatic multi-mode adaptation to various RF inputs by sensing RF input voltage and configuring the B^2R^3 accordingly. A conventional Dickson charge pump rectifier (Fig. 6.8) works as an envelope detector, which provides information on RF input voltage to following two comparators. With two references voltages, REF_1 and REF_2, the two comparator and digital logic produce outputs for an appropriate configuration. The comparator having three output states for the mode arbiter is shown in Fig. 6.5(C). I_B is set to 40 nA and total static current for the comparator is 200 nA from 0.8 V supply.

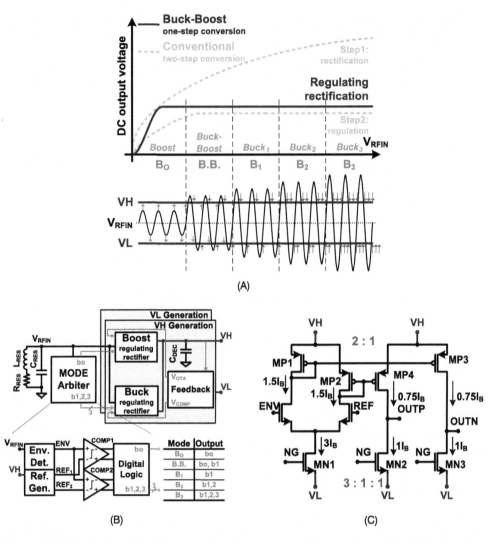

Figure 6.5 Methodology of multi-mode adaptation to various RF inputs is detailed. (A) Mode switching among 5 different buck and boost regulating rectification modes supports wide RF input range. (B) Mode arbiter senses RF input voltage and changes configuration of the B^2R^3 automatically with 4 digital output bits, bo, b1–b3. (C) The simplified schematic of the proposed comparator having three output states for the mode-arbiter is shown.

6.5.2 Feedback

Feedback is essential for regulation. The feedback module first senses output voltage and delivers feedback signal to either a boost RR or a buck RR to adjust effective R_{ON} resistance between V_{RFIN} and VH or VL for regulation. Two diode–connected MOSFETs, R_S, in parallel with a capacitor, C_S, shown in Fig. 6.6(A) senses output

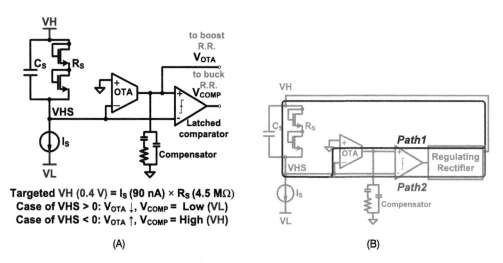

Figure 6.6 (A) Block diagram of the dual path feedback module, and (B) dual path feedback loop for fast and accurate regulation.

voltage, VH for VH generation. Owing to voltage drop from R_S and I_S, VHS is simply DC level shifted from VH and is compared to ground at the input of a OTA and a latched comparator. I_S is set to 90 nA while R_S is 4.5 MΩ for 0.4 V of targeted VH. To avoid significant RC delay from R_S and parasitic capacitance at a VHS node, 150 fF capacitance, C_S, is added in parallel with the two diode connected MOSFETs.

The simplest modality for adjustment of an effective R_{ON} resistance is an on/off control, so-called bang-bang control, which turns off a regulating rectifier (RR) if output voltage is greater than a target voltage, and vice versa. Bang-bang control features simple implementation and fast load regulation since it requires only one comparator and no delay components for compensation. However, it has unavoidable disadvantage: output voltage is never able to reach a target voltage. In other words, bang-bang control always permits a certain amount of error between output and target voltages. On the other hand, an error integration control that continuously adjusts R_{ON} resistance of RR directly based on information of integrated error. Although this error integration control eliminates error completely at the output voltage nodes, it induces lagging on feedback signal. As such, the speed of load regulation is limited. The B^2R^3 features a dual feedback path for a combination of an error integration control (Path 1) and a bang-bang control (Path 2), enabling both fast load regulation and accurate control on the output voltage as shown in Fig. 6.6(B). A conventional folded cascode amplifier is employed for the OTA and consumes 400 nA of static current from 0.8 V supply. The latched comparator consists of a pre-amplifying stage and a latch stage. Pre-amplifying stage has three cascaded self-biased inverter type amplifiers [31] followed by a conventional inverter type clocked-latch. The latched comparator consumes 600 nA of static

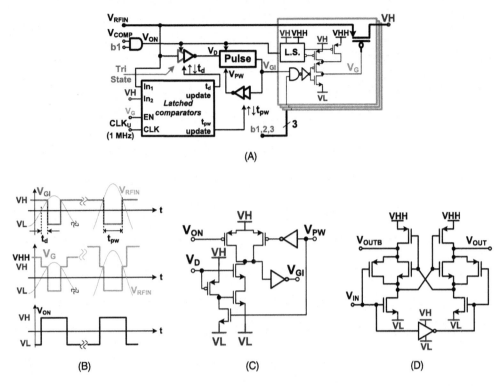

Figure 6.7 (A) Block diagram of the buck-mode regulating rectifier, and (B) time waveforms. Detailed circuit schematics for (C) pulse generator and (D) level-shifter in the buck-mode regulating rectifier.

current and 800 nA dynamic current at 32 MHz clock speed for the latch from 0.8 V supply. Pre-amplifying stage amplifies difference between V_{OTA} and VHS to CMOS level (rail-to-rail) while it induces less than 10 ns delay.

6.5.3 Buck-mode RR

When RF input amplitude is greater than 0.4 V, the mode arbiter configures the B^2R^3 to buck modes by asserting b1 to permit a buck-mode regulating rectifier (buck RR) shown in Fig. 6.7(A) to be enabled according to V_{COMP}. The enabled buck RR (V_{COMP}: HIGH) basically implements a conductive path from RF input, V_{RFIN}, to VH through a power PMOS when V_{RFIN} is greater than VH. This path is power efficient owing to low R_{ON} resistance of the PMOS since the PMOS is fully turned on by lowering V_G to VL. However, because of the low R_{ON} resistance, turning on/off the PMOS at the optimal time is critical. Otherwise, power conversion efficiency (PCE) of the RR is severely degraded due to either huge reverse current for the case of wider on-time or switching loss coming from gate capacitance of the large sized power PMOS for the case of narrow on-time. Typically, at the expense of significant power consumption, high speed comparators have been employed hitherto to turn on the PMOS when V_{RFIN} starts to greater than VH and turn off when V_{RFIN} becomes less than VH, which is the optimal time for PCE [32]. Although a recent work presents a power-efficient high speed comparator [33], it requires additional adjustable delay element to turn off the PMOS. It turns out that if local negative feedback loops figure out only two time variables, t_d and t_{pw} in Fig. 6.7(B), the high-speed comparators are no longer required [7,23]. These time variables are updated every 1 μs by comparing V_{RFIN} and VH with two conventional latched comparators at falling and rising edge of the gate voltage, V_G, for each variable to maximize t_{pw} for the optimal on-time. Since a current starved inverter chain is utilized for implementing each time variable, V_D and V_{PW} have slow rising and falling time, possibly inducing non-negligible short-circuit current with a conventional logic for pulse generation. As such, a pulse generator that has three inputs, V_D, V_{PW} and V_{ON}, one output, V_{GI}, and offers no short-circuit current is presented as shown in Fig. 6.7(C).

On the other hand, when the buck RR is disabled, it is important to turn off the PMOS completely. If the PMOS is gated with the voltage level of VH, whenever V_{RFIN} is greater than V_{TH} on top of VH, a weak conductive path from V_{RFIN} to VH is induced through the PMOS. As such, it fails to regulation at high RF input case. Furthermore, this path has large R_{ON} resistance, leading to severe degradation in PCE. Hence, the PMOS should be gated with higher voltage level than V_{RFIN}. For this purpose, VHH is internally generated greater than V_{RFIN} and utilized to gate the PMOS completely with a level-shifter shown in Fig. 6.7(D). This presented level-shifter consumes 1.5 μW at 10 MHz switching frequency with 100 fF capacitive load for less than 1.5 ns rising time, which is a 3× improvement over a conventional design.

6.5.4 Boost-mode RR

Even if RF input amplitude is less than target DC voltage, RF to DC rectification is feasible with a boost-mode rectifier. A conventional Dickson charge-pump has been widely adopted for boost-mode RF to DC rectification [13] because of its simplicity for implementation at the expense of losses in PCE and VCE from V_{TH} of two diode-connected NMOSs. It is viable to alleviate V_{TH} losses by inserting a floating voltage source between gate and drain of NMOS as depicted in Fig. 6.8(B) since positive voltage V_{OTA} from the inserted voltage source cancels V_{TH} voltage of the NMOS. Here, regulation functionality is simply added along with rectification by adjusting V_{OTA} to adjust the amount of the V_{TH} cancellation as shown in Fig. 6.8(C). The amount of the V_{TH} cancellation should be adjusted by a feedback module monitoring a output voltage. Luckily, the B^{2}R^3 has the feedback module for the buck RR. Hence, no additional feedback module is required for the boost-mode regulating rectifier.

Fig. 6.9 is a block diagram of the boost RR utilizing switched capacitor circuits as floating voltage sources with high-voltage non-overlapping clocks. A conventional non-overlapping clock generator [34] provides non-overlapping clock out of a single-

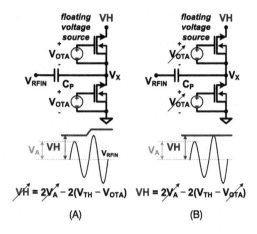

Figure 6.8 Boost RF-DC rectifiers; (A) a V_{TH} cancellation rectifier, and (B) a V_{TH} cancellation regulating rectifier.

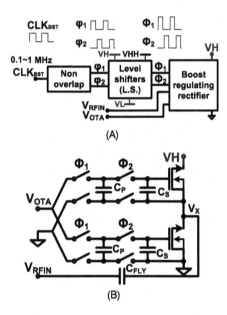

Figure 6.9 Implementation of the boost-mode regulating rectifier; (A) block diagram with clock time waveforms, and (B) a V_{TH} cancellation regulating rectifier with switched capacitor circuits serving as floating voltage sources.

ended clock ranging from 0.1 to 1 MHz, and following level shifters change voltage domain from VH to VHH. Non-overlapping clock generator and level-shifters consume 200 nW at 1 MHz clock speed from 0.8 V VH supply and 1.2 V VHH supply. C_P, C_S, and C_{FLY} in Fig. 6.9(B) are 1.2, 3, and 22 pF, respectively.

6.6. Measurement results

The fully integrated wireless-power-receiver-on-chip (WiPow-RX) including the B^2R^3 were measured with an external 23 mm × 23 mm coil shown in Fig. 6.10(A). Quality factor (Q) of the TX coil is 160 at 144 MHz. This TX coil is driven by a power amplifier (Mini-Circuits, ZHL-42W+). The 3 mm × 3 mm WiPow-RX in Fig. 6.10(B) is powered through inductive coupling from the TX coil. The B^2R^3 provides dual regulated power supplies, VH and VL, to drive on-chip loads including amplifiers and an amplitude-shift-keying (ASK) module. The RX coil is 3 turn for 60.3 nH inductance, L_{RES}. The Q factor of the RX coil is measured as 10.9 at 144 MHz, rendering series parasitic resistance R_{RES} to be 5 Ω. This series resistance is equivalent to 600 Ω of parallel resistance (R_P) by series-to-parallel conversion. Simulated (ANSYS HFSS) L_{RES}, Q factor, and R_P of the RX coil are 60 nH, 11.2, and 610 Ω, respectively. The recent work details methodology of inductive link optimization and its results [35]. Measurement characterization was done with 10 mm of link distance between TX and RX coils. Otherwise, link distance is specified.

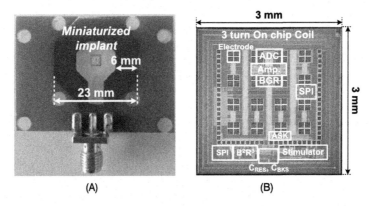

Figure 6.10 (A) External 23 mm × 23 mm octagonal power TX antenna, and (B) microphotograph for fully integrated wireless-power-receiver-on-chip (WiPow-RX) with low loss H-tree distribution network.

Under RF input envelope variation induced by any changes of link environment, the presented B^2R^3 performs input regulation through mode-adaptation for coarse regulation and dual path feedback for fine regulation as shown in Fig. 6.11. The primary coil was driven by the power amplifier receiving amplitude modulated signals to generate RF input envelope transient variation, as depicted in blue-gray colored background waveform. The mode arbiter in the B^2R^3 sensed RF input envelope and changed configuration of the B^2R^3 adaptively from buck1 mode to buck3 mode for (A) and from buck-boost mode to buck2 mode for (B). Independently, the feedback module also sensed output voltages and provided a feedback signal to each regulating rectifier.

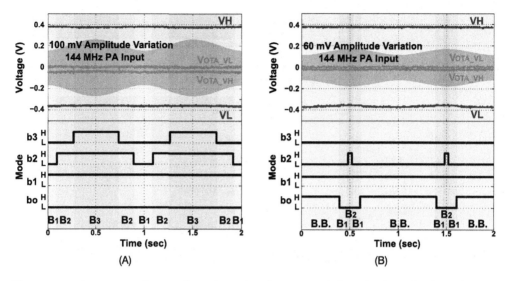

Figure 6.11 RF input regulations through mode adaptation; (A) from buck1 to buck3 mode, and (B) from buck-boost to buck2 mode based on RF input envelope.

Owing to these combined regulation methodologies, the B^2R^3 offers rectification and regulation under RF input variation.

Load regulation is an important function for the B^2R^3 since on-chip loads such as ADC and stimulator draw switching current out of supply voltages. To mimic radical load variation, external time-varying current load from 0 to 200 μA was connected between VH and VL, and output voltages VH and VL were measured as shown in Fig. 6.12(A). Other than on-chip 250 pF decoupling capacitance implemented with the H-tree (two 500 pF capacitors in series for VH-GND and GND-VL), no additional decoupling capacitor was connected between output nodes. The buck-mode regulating rectifier in the B^2R^3 performed rectification and regulation with the dual path feedback. When load is heavy, ripple voltage at output nodes is decreased to 3 mV because output nodes had smaller resistance than resistance with light load. Since the dual path feedback features fast regulation owing to a bang-bang control path (path 2, Fig. 6.6), the B^2R^3 offered negligible undershoot and overshoot as compared a prior art utilizing only an error integration feedback path similar to path 1 of Fig. 6.6(B) [33].

Transfer function from the input of the power amplifier (PA) to output voltages (V_{OUT}) of the B^2R^3 in Fig. 6.13 was measured by changing input with three different load conditions at 10 mm of link distance between TX coil to RX on-chip coil. At small RF input, the B^2R^3 operated in the boost-mode to generate target output voltage while at large RF input, high voltage gating technique in the buck-mode regulating rectifier blocked an unintended and inefficient conductive path generated by a power

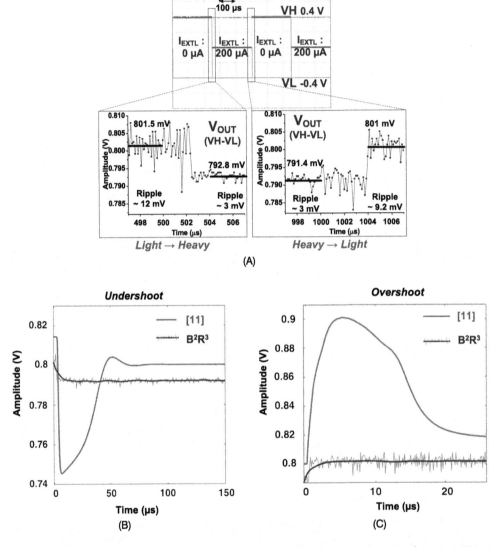

Figure 6.12 (A) Load regulation measurement with external load current changing from 0 to 200 μA to 0.8 V V_{OUT}. (B) Undershoot and (C) overshoot performance comparison with a prior art [33].

PMOS diode-connection from RF input to output voltages to maintain target voltage as explained in Sects. 6.5.3 and 6.5.4. Hence, the B^2R^3 has 11 dB of the wide RF input range compared to a prior-art that has 4 dB RF input range [33]. Greater than −3 dB RF input may reach breakdown voltage of MOSFET transistors in gate drivers of the buck RRs.

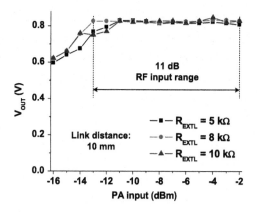

Figure 6.13 Transfer function curve showing RF input range of the B^2R^3.

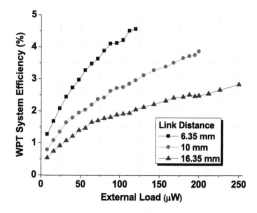

Figure 6.14 Measured overall wireless power transmission (WPT) system efficiency (WSE) from TX to regulated V_{OUT}.

Overall WSE shown in Fig. 6.14 was measured at three different link distances. Power losses from the TX matching network, PCB traces, SMA connectors, and cables were measured through a network analyzer (Keysight, E5080A) and de-embedded to obtain the accurate power gain [16,36,33]. Interestingly, the B^2R^3 drive heavier loads at farther link distance since at closer distance, the induced RF input voltage at the secondary LC tank V_{RFIN} with the same RF transmitted power becomes larger enough to turn on electrostatic discharge (ESD) diodes consuming the transferred power. As such, the B^2R^3 produces lower supplies than target voltages to heavier loads after ESD diodes are enabled, even if the transmitted power is increased. It is worth noting that, compared to the recent wireless power receiver with two-turn on-chip RX coil [33], the presented WiPox-RX receives RF energy with greater efficiency owing to (i) better impedance matching between the secondary LC tank and the B^2R^3 owing to relatively large impedance of the secondary LC tank at resonance, 600 Ω, and (ii) better quality factor of the RX coil maximized by low-loss H-tree power and signal distribution network. Measured overall WSE reaches up to 4.5%, 3.9%, and 2.8% at 6.35, 10, and 16.35 mm link distance, respectively.

For system level validation, measurements for an analog front-end (AFE) module and an amplitude-shift keying (ASK) module along with the B^2R^3 were conducted. AFE is the most sensitive analog building block in the miniaturized implant since it amplifies small electrocortical signal ($\ll 1$ mV) to a few hundred mV for easy digitization as shown in Fig. 6.15. Due to this small amplitude of input signal, the presented H-tree should de-

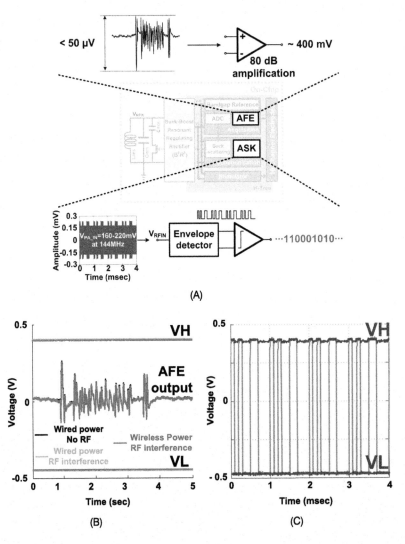

(A)

(B)

(C)

Figure 6.15 System level validations; (A) test set-up, (B) analog front-end output, and (C) ASK output.

couple RF interference and convert residual RF interference to common-mode noise. Given the same input, Fig. 6.15(B) shows AFE output voltages with three different conditions at; wired power supplies without RF interference (shown in black), wired power

Table 6.1 Performance comparison.

	Muller JSSC 2015	Driscoll ISSCC 2009	Mark VLSI 2011	Zargham TBioCAS 2015	Kim JSSC 2017	This work
Resonant frequency (MHz)	300	915	535	160	144	144
RX Coil /Value(nH) /Area(mm^2)	Off-chip /32 /42.25	Off-chip /N/R /4	Off-chip /5.73 /1	On-chip /130 /4.36	On-chip /23.7 /8.64	On-chip /60.3 /8.74
Number of modes	1	1	1	1	1	5
Regulation approach	Separate LDO	Separate LDO	Separate LDO	Separate LDO	Regulating rectifier	Regulating rectifier
Overshoot (mV)	N/R	N/R	N/R	N/R	100	< 1
Decoupling capacitor (nF)	4	N/R	1.39	0.02	1	0.25
Process	65 nm CMOS	0.13 um CMOS	65 nm CMOS	0.13 µm CMOS	0.18 µm CMOS SOI	0.18 µm CMOS SOI
Overall WSE (%)	1.19[a,c]	0.05 (−33 dB)	0.02 (−37 dB)	0.62[b]	2.04[c]	3.39[c]
Distance (mm)	12.5	15	13	10	10	10
WPT system efficiency FoM*	8.46	20.25	43.94	68.1	80.3	131.2

[a] Estimated from provided data; TX PW: 13 mW and P_{dc_load}: 160 µW.
[b] Estimated from provided data, estimated η_{LDO}: 68% (V_{DD}: 3.1 V, V_{REC2}: 4.5 V), provided η from TX coil to output of the rectifier: 0.9%.
[c] With an external load of 160 µW.

* WPT system efficiency FoM $= \dfrac{\eta_{overall} \times D^3}{A^{1.5}}$, where $\eta_{overall}$ is TX-regulated DC efficiency; D is distance between TX-RX coils; and A is area of the RX coil.

supplies with RF interference (red), and wireless power transmission (WPT) with RF interference (blue). Owing to the efficacy of the H-tree, the amplified waveforms for three different cases are hardly distinguishable. Although ASK is the energy-efficient data communication methodology, it produces severe disturbance to the B^2R^3 since the ASK module inherently utilize amplitude variation at received RF input voltage as shown in Fig. 6.15(A). The measured input voltage to the power amplifier and output voltage of the ASK module along with regulated DC voltages VH and VL under WPT case are shown in Fig. 6.15(C).

Table 6.1 summarizes key performance measures of the B^2R^3 in comparison to state-of-the-art designs for mm-sized implants. Overall WSE is significantly improved with simultaneous regulating rectification. Owing to dual-path feedback, regulation is fast enough to have negligible over/undershoot while keeping high loop gain for better regulation performance. For fair comparison, the figure-of-merit (FoM) [33] considering that for mm-sized RX coils (in the mid-field regime) the TX-RX link efficiency is inversely proportional to the cube of the inter-coil distance [18] and the cube of square root of RX coil area [21] is adopted.

6.7. Conclusions

Miniaturization of implants enables long-term and robust BMI technologies by improving their longevity, safety, and high spatial resolution [37]. This review highlighted two major WPT methods for mm-sized implants: ultrasonic WPT and electromagnetic WPT. While ultrasonic WPT offers advantages particularly for sub-mm implants, electromagnetic WPT is superior for BMI applications, as it is better suited for transcranial transmission and can support higher data rates. Critical design considerations for optimal integration of RX coils on-chip include sizing and number of turns of the coil, and H-tree power and signal distribution. Two example designs for regulating rectifiers operating as fully integrated wireless power receivers were presented with high WPT system efficiency figure-of-merit. As a significant advancement in miniaturization, fully integrated wireless power receivers enable next-generation modular mm-sized wireless implants. On-going and future research directions include closed-loop communication between external beamforming transceiver and distributed modular implants [38].

References

[1] M. Capogrosso, T. Milekovic, D. Borton, F. Wagner, E.M. Moraud, J.-B. Mignardot, N. Buse, J. Gandar, Q. Barraud, D. Xing, E. Rey, S. Duis, Y. Jianzhong, W.K.D. Ko, Q. Li, P. Detemple, T. Denison, S. Micera, E. Bezard, J. Bloch, G. Courtine, A brain-spine interface alleviating gait deficits after spinal cord injury in primates, Nature 539 (7628) (2016) 284–288.

[2] S.S. Polikov, P.A. Tresco, W.M. Reichert, Response of brain tissue to chronically implanted neural electrodes, Journal of Neuroscience Methods 148 (1) (2005) 1–18.

[3] H. Lee, R.V. Bellamkonda, W. Sun, M.E. Levenston, Biomechanical analysis of silicon microelectrode-induced strain in the brain, Journal of Neural Engineering 2 (4) (Dec. 2005) 81–89.

[4] G.C. McConnell, H.D. Rees, A.I. Levey, C.-A. Gutekunst, R.E. Gross, R.V. Bellamkonda, Implanted neural electrodes cause chronic, local inflammation that is correlated with local neurodegeneration, Journal of Neural Engineering 6 (5) (Aug. 2009).

[5] L. Karumbaiah, T. Saxena, D. Carlson, K. Patil, R. Patkar, E.A. Gaupp, M. Betancur, G.B. Stanley, L. Carin, R.V. Bellamkonda, Relationship between intracortical electrode design and chronic recording function, Biomaterials 34 (33) (Nov. 2013) 8061–8074.

[6] S. Ha, A. Akinin, J. Park, C. Kim, H. Wang, C. Maier, P.P. Mercier, G. Cauwenberghs, Silicon-integrated high-density electrocortical interfaces, Proceedings of the IEEE 99 (2016) 1–23.

[7] R. Muller, H.-P. Le, W. Li, P. Ledochowitsch, S. Gambini, T. Bjorninen, A. Koralek, J. Carmena, M. Maharbiz, E. Alon, J. Rabaey, A minimally invasive 64-channel wireless μECoG implant, IEEE Journal of Solid-State Circuits 50 (1) (Jan. 2015) 344–359.

[8] H.-M. Lee, M. Ghovanloo, A high frequency active voltage doubler in standard CMOS using offset-controlled comparators for inductive power transmission, IEEE Transactions on Biomedical Circuits and Systems 7 (3) (Jun. 2013) 213–224.

[9] S. Ozeri, D. Shmilovitz, Ultrasonic transcutaneous energy transfer for powering implanted devices, Ultrasonics 50 (6) (2010) 556–566.

[10] M.D. Menz, Ã. Oralkan, P.T. Khuri-Yakub, S.A. Baccus, Precise neural stimulation in the retina using focused ultrasound, Journal of Neuroscience 33 (10) (2013) 4550–4560.

[11] D. Seo, J.M. Carmena, J.M. Rabaey, M.M. Maharbiz, E. Alon, Model validation of untethered, ultrasonic neural dust motes for cortical recording, Journal of Neuroscience Methods 244 (2015) 114–122.

[12] J. Charthad, M.J. Weber, T.C. Chang, A. Arbabian, A mm-sized implantable medical device (IMD) with ultrasonic power transfer and a hybrid bi-directional data link, IEEE Journal of Solid-State Circuits 50 (8) (Aug. 2015) 1741–1753.

[13] M. Mark, Powering mm-Size Wireless Implants for Brain–Machine Interfaces, Ph.D. dissertation, Electrical Engineering and Computer Sciences, University of California at Berkeley, Dec. 2011.

[14] P.N.T. Wells, Biomedical Ultrasonics, Academic Press, 1977.

[15] E.E. Christensen, T.S. Curry III, J.E. Dowdey, An Introduction to the Physics of Diagnostic Radiology, second edition, Lea & Febiger, 1978.

[16] D. Ahn, M. Ghovanloo, Optimal design of wireless power transmission links for millimeter-sized biomedical implants, IEEE Transactions on Biomedical Circuits and Systems 10 (1) (Feb. 2016) 125–137.

[17] S. O'Driscoll, A. Poon, T. Meng, A mm-sized implantable power receiver with adaptive link compensation, in: IEEE International Solid-State Circuits Conference Digest of Technical Papers (ISSCC), Feb. 2009, pp. 294–295.

[18] A. Poon, S. O'Driscoll, T. Meng, Optimal frequency for wireless power transmission into dispersive tissue, IEEE Transactions on Antennas and Propagation 58 (5) (May 2010) 1739–1750.

[19] M. Mark, Y. Chen, C. Sutardja, C. Tang, S. Gowda, M. Wagner, D. Werthimer, J. Rabaey, A 1 mm^3 2 Mbps 330 fJ/b transponder for implanted neural sensors, in: Symposium on VLSI Circuits (VLSI Circuits), Jun. 2011, pp. 168–169.

[20] M. Zargham, P.G. Gulak, A 0.13 μm CMOS integrated wireless power receiver for biomedical applications, in: Proceedings of the ESSCIRC (ESSCIRC), Sept. 2013, pp. 137–140.

[21] M. Zargham, P. Gulak, Fully integrated on-chip coil in 0.13 μm CMOS for wireless power transfer through biological media, IEEE Transactions on Biomedical Circuits and Systems 9 (2) (Apr. 2015) 259–271.

[22] C. Kim, S. Ha, J. Park, A. Akinin, P. Mercier, G. Cauwenberghs, A 144 MHz integrated resonant regulating rectifier with hybrid pulse modulation, in: Symposium on VLSI Circuits (VLSI Circuits), June 2015, pp. 284–285.

[23] C. Kim, J. Park, A. Akinin, S. Ha, R. Kubendran, H. Wang, P. Mercier, G. Cauwenberghs, A fully integrated 144 MHz wireless-power-receiver-on-chip with an adaptive buck-boost regulating rectifier and low-loss H-Tree signal distribution, in: Symposium on VLSI Circuits (VLSI Circuits), June 2016.

[24] R. Sarpeshkar, Ultra Low Power Bioelectronics, Cambridge University Press, 2010.

[25] P.P. Mercier, A.P. Chandrakasan, Ultra-Low-Power Short-Range Radios, Springer, 2015.

[26] S. Kim, J.S. Ho, A.S.Y. Poon, Wireless power transfer to miniature implants: transmitter optimization, IEEE Transactions on Antennas and Propagation 60 (10) (Oct. 2012) 4838–4845.

[27] B. Lee, M. Kiani, M. Ghovanloo, A triple-loop inductive power transmission system for biomedical applications, IEEE Transactions on Biomedical Circuits and Systems 10 (1) (Feb. 2016) 138–148.

[28] X. Li, C.Y. Tsui, W.H. Ki, A 13.56 MHz wireless power transfer system with reconfigurable resonant regulating rectifier and wireless power control for implantable medical devices, IEEE Journal of Solid-State Circuits 50 (4) (Apr. 2015) 978–989.

[29] L. Cheng, W.H. Ki, C.Y. Tsui, A 6.78-MHz single-stage wireless power receiver using a 3-mode reconfigurable resonant regulating rectifier, IEEE Journal of Solid-State Circuits 52 (5) (May 2017) 1412–1423.

[30] H.M. Lee, C.S. Juvekar, J. Kwong, A.P. Chandrakasan, A nonvolatile flip-flop-enabled cryptographic wireless authentication tag with per-query key update and power-glitch attack countermeasures, IEEE Journal of Solid-State Circuits 52 (1) (Jan. 2017) 272–283.

[31] M. Bazes, Two novel fully complementary self-biased CMOS differential amplifiers, IEEE Journal of Solid-State Circuits 26 (2) (Feb. 1991) 165–168.

[32] H.-M. Lee, M. Ghovanloo, An integrated power-efficient active rectifier with offset-controlled high speed comparators for inductively powered applications, IEEE Transactions on Circuits and Systems I: Regular Papers 58 (8) (Aug. 2011) 1749–1760.

[33] C. Kim, S. Ha, J. Park, A. Akinin, P. Mercier, G. Cauwenberghs, A 144 MHz fully integrated resonant regulating rectifier with hybrid pulse modulation for mm-sized implants, IEEE Journal of Solid-State Circuits (2017).

[34] D.A. Johns, K. Martin, Analog Integrated Circuit Design, Wiley, 1996.

[35] J. Park, C. Kim, A. Akinin, S. Ha, G. Cauwenberghs, P.P. Mercier, Wireless powering of mm-scale fully-on-chip neural interfaces, in: IEEE Biomedical Circuits and Systems Conference (BioCAS) Proceedings, Oct. 2017.

[36] A. Ibrahim, M. Kiani, A figure-of-merit for design and optimization of inductive power transmission links for millimeter-sized biomedical implants, IEEE Transactions on Biomedical Circuits and Systems 10 (6) (Dec. 2016) 1100–1111.

[37] C. Kim, S. Ha, A. Akinin, J. Park, R. Kubendran, H. Wang, P.P. Mercier, G. Cauwenberghs, Design of miniaturized wireless power receivers for mm-sized implants, in: 2017 IEEE Custom Integrated Circuits Conference (CICC), Austin, TX, May 2017.

[38] C. Kim, S. Joshi, C. Thomas, S. Ha, L. Larson, G. Cauwenberghs, A 1.3 mW 48 MHz 4-channel MIMO baseband receiver with 65 dB harmonic rejection and 48.5 dB spatial signal separation, IEEE Journal of Solid-State Circuits 51 (4) (Apr. 2016) 832–844.

CHAPTER 7

Wireless data communication for ECoG implants

Contents

7.1. Challenges of ECoG implants in wireless data communication

The proliferation of power and form-factor constrained devices like implanted medical devices (IMDs), radio-frequency identification (RFID) and wireless sensors has led to the extensive use of inductive wireless links for short-range power and data telemetry. For these applications including IMDs with an implantation depth of a few cm, such as cochlear implants, visual prostheses and brain–computer interface (BCI) systems with high-throughput neural recording, wireless inductive powering is a natural choice as the primary mode of power transfer due to its high efficiency and well known robustness in comparison to competing technologies such as ultrasound telemetry and energy scavenging. Thus, over the past decades, inductive power telemetry has been the focus of extensive studies resulting in the development of many efficient designs and methodologies [1–8].

Since near-field communication (NFC) using inductive links offers lower cost of communication than far-field communication methods, it has become the primary means of communication with IMDs [9–15] and sensors [16–21], and further has been adopted as a secondary communication channel between power-constrained mobile devices such as smartphones and tablets. Along with NFC, RFID applications also use inductive links to power battery-less tags and sensors while transferring stored data [22, 23].

In both biomedical and sensor applications, available power and geometry on the primary and secondary sides are very asymmetric. Implanted devices and sensor nodes have much more stringent constraints both in power consumption and size while those constraints are significantly more relaxed on the primary side [7]. Thus, both uplink and downlink data telemetry need to be substantially more power and area efficient on the secondary side than on the primary side. Hence, transmitters on the implanted or sensor side for uplink telemetry require more stringent optimization for efficiency than receivers on the external side. In addition, simpler antennas are preferred on the implanted and sensor side due to size and reliability constraints. While several power efficient schemes for high-data-rate forward telemetry [24–27,10,28] have been demonstrated for implants and sensor nodes, efficient high-data-rate backward telemetry in these settings has remained challenging.

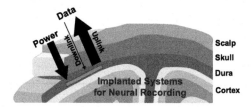

Figure 7.1 Simplified depiction of implanted neural recording systems and their requirements of power and data telemetry.

To highlight the challenges in the design of such systems, we specify the requirements that must be met by a typical high-density brain activity monitoring IMD (see Fig. 7.1). These IMDs typically have one of three configurations: an entire device placed on the cortex [12]; an electrode array placed on the cortex with the primary device on the craniotomy [29]; or the primary device placed under the scalp [30,31]. In all these configurations, the devices are highly geometry and power constrained and the implant location poses a major challenge of power acquisition and data communication (Fig. 7.1). The data rate requirement for 1024-channel ECoG recording with a sampling rate of 600 S/s and a data resolution of 10 bits is 6.15 Mbps. Similarly, for 64-channel recording of neural spikes and local field potentials at 10-kS/s rate and 10-bit resolution, a data rate of 6.4 Mbps is required. State-of-the-art IMDs recording and transmitting data at these high rates typically consume several mW of power [32–37,12,29]. Since the antenna geometry and ASIC placement near the implant site constrain available power, achieving both the power transfer and data rate specifications proves to be a challenging task.

7.2. General approaches

To meet these requirements, one popular approach is to use multiple dedicated inductive links for power and data telemetry as shown in Fig. 7.2(A) [38,39,25]. Since each link can be optimized independently, this approach allows achieving high data rate

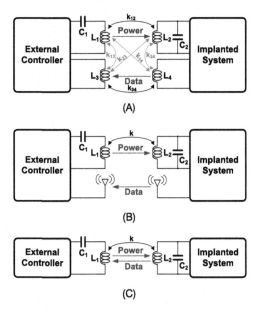

Figure 7.2 Power and uplink data telemetry topologies: (A) separate inductive links for power and data telemetry, (B) inductive link for power transfer and active radio for data telemetry, and (C) single inductive link.

[39,40,27,41] while maintaining high power transfer efficiency (PTE). However, this approach requires a more complicated antenna structure and suffers from cross-talk between the links [38,25,31,42].

Active transmission using higher radio frequency (RF) bands as in Fig. 7.2(B) can also be adopted. This scheme can easily achieve the required data rate. However, it consumes an order of magnitude more energy and incurs the cost of increased complexity in circuit and antenna structures. Recently, various carrier-less pulse-based transmission technologies such as impulse radio ultra-wideband (IR-UWB) have been developed for short range communication between low-power sensors [33,43–45]. These transmitters do not need high precision frequency/phase reference generation, resulting in lowered power consumption and greater amenability for adoption in wireless sensors. However, strong attenuation by electromagnetic absorption in the body at UWB bands from 3.1–10.6 GHz reduces its effectiveness in implantable applications.

The simplest approach uses a single inductive link to transfer both power and data as depicted in Fig. 7.2(C). Typically, back scattering is used for passive backward telemetry. By not actively driving the antenna, power consumption on the implanted side can be minimized. One of the most widely used back-scattering schemes is load shift keying (LSK), which modulates the load on the secondary side by shorting and/or opening the secondary LC tank [46,39,47]. A major drawback of LSK over a single inductive link is that the conditions for efficient power transmission and high data rate are incongruent:

PTE requires a high quality factor Q of the inductive link, at the expense of reduced data rate [7]. In particular, the maximum data rate with LSK is inversely proportional to the quality factor as $\sqrt{2}(f_{carrier}/Q)$ [48]. With simultaneous power transfer, achievable data rates have been limited to 100–500 kbps over a 13.56-MHz inductive link [49,50].

Moreover, there are two often-neglected sources of power loss in LSK. Shorting the tank depletes all the stored energy. Also, while the tank is shorted, the rectifier does not receive power from the coil, leading to a reduction of delivered power to the load. These two aspects are not typically accounted for in the reported power consumption of telemetry ICs.

Recently increased data-rates were achieved in [51] by using the transient response from passive phase shifts by shorting the secondary LC tank for a half-cycle, resulting in 0.858 Mbps with simultaneous power transfer over a single inductive link. However, this scheme suffers from a double loss of tank energy because of complete phase reversal of LC resonance, and still is subject to the trade-off between PTE and data rate due to resonance recovery time after each bit transmission.

The increasing density and bandwidth requirements from emerging IMDs and sensors motivate development of alternative modulation schemes that obviate the PTE and data rate trade-off. To this end we present a new back-scattering data modulation method over a single inductive link that achieves high data rate at low power consumption while not compromising PTE, by recycling most of the energy circulating in the resonant tank during data transmission. Initial results reported in [52] demonstrated transmission of two data bits for every four carrier cycles while minimally depleting the resonant tank energy. For neural recording implants, forward (downlink) telemetry requires a very low data rate with infrequent transmission for sending configuration bits. Thus, such forward telemetry can share the same inductive link by time multiplexing, or very slow amplitude modulation can be used simultaneously for full duplex communication.

7.3. COOK modulation scheme

7.3.1 Operation principle

The proposed uplink data modulation scheme, *cyclic on–off keying* (COOK), utilizes a shorting switch across the secondary LC tank to modulate data in a manner similar to conventional LSK, but with some fundamental differences. Whereas LSK modulation typically closes the switch for tens to hundreds of cycles, COOK modulation closes the switch for a single cycle only. Fig. 7.3 shows the operation of the circuit when the modulation switch is (A) open (DATA = 0), and (B) closed (DATA = 1). Unlike LSK closing and opening the switch across the secondary LC tank at random phases, COOK starts closing the switch when the voltage across the secondary inductor L_2 is zero and the current through the inductor I_{L2} is at an extremum. At this time in-

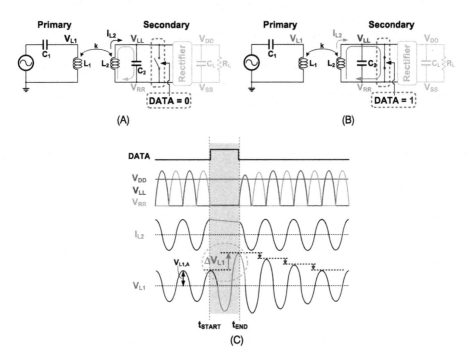

Figure 7.3 Conceptual illustration of the proposed cyclic on–off keying (COOK) modulation scheme. (A) The secondary LC tank is at resonance when the switch is open (DATA = 0), cycling energy between L_2 and C_2 in the form of quadrature current and voltage. (B) Closing the switch (DATA = 1) when the secondary tank voltage reaches zero recirculates all resonant energy as DC current through the loop formed by L_2 and the switch, while maintaining zero voltage across the tank. Reopening the switch after a complete cycle reestablishes the initial conditions for the secondary tank resonance. (C) Voltage and current waveforms throughout the modulation operation.

stance (t_{START} in Fig. 7.3(C)), all the energy in the secondary LC tank is in the current through the inductor. During the synchronized shorting cycle (from t_{START} through t_{END} in Fig. 7.3(C)), the current through the inductor and the switch remains constant, circulating and maintaining all energy (except for energy loss through parasitic resistances R_2 and R_{sw} of the secondary inductor and the modulation switch) as shown in Fig. 7.3(B). The instantaneous disruption of the inductive coupling to the primary side results in a voltage change at V_{L1}, indicated as ΔV_{L1} in Fig. 7.3(C).

The COOK synchronized shorting does not disturb resonance and thus, unlike most other data telemetry schemes, does not compromise the high Q of the secondary coil for resonant power transfer. Unlike LSK, the shorting does not dissipate energy in the secondary LC tank except for losses due to the parasitic resistances of the inductor and the switch. Shown in Fig. 7.3(C), the secondary coil voltages V_{LL} and V_{RR} resume their normal course immediately after the COOK shorting and recover to the peak resonance amplitudes within a few cycles. In addition, a large fraction of the secondary coil energy is preserved after synchronized shorting, in contrast to LSK and other conventional schemes where this energy loss is not accounted for in the calculation of transmission energy per bit.

On the primary side, a voltage rise is immediately induced during the short of the secondary coil as shown on the bottom panel of Fig. 7.3(C). Under typical under-coupled conditions of the inductive link, the primary coil voltage amplitude exponentially decreases to baseline after the transient increase during the short, until the next shorting cycle. Hence a rise in the peak of the primary coil voltage V_{L1} occurs only during a shorting cycle, and transmitted data can be decoded directly from the consecutive-cycle peak differentials ΔV_{L1} as indicated in Fig. 7.3(C). By locking on to the phase of V_{L1} on the primary side, peak detection circuitry for data decoding is greatly simplified.

7.3.2 Dependence on primary and secondary quality factors

Fig. 7.4 compares simulated resonance recovery times for COOK and conventional LSK after shorting of the secondary coil switch. Recovery time for LSK modulation increases

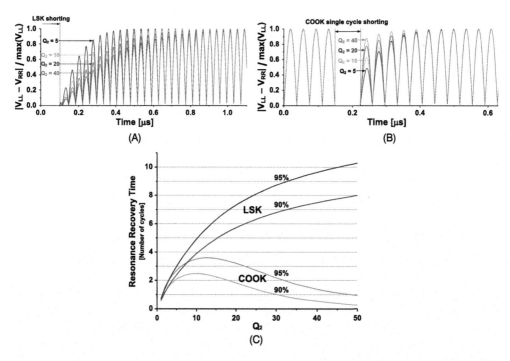

Figure 7.4 Simulated transient voltage waveforms of the secondary coil after shorting with (A) LSK and (B) COOK modulation. (C) Recovery time of tank resonance after shorting for LSK and COOK modulation to reach 90% and 95% of the original voltage amplitude.

with increasing Q_2 as shown in Fig. 7.4(A). Hence, achievable data rate with LSK is inversely proportional to Q_2. On the contrary, recovery time for COOK modulation shown in Fig. 7.4(B) is shorter, and even decreases for Q_2 beyond 10 as shown in Fig. 7.4(C). The decreasing recovery time despite longer time constant of the resonant tank at higher Q_2 is mainly due to a reduced amplitude droop in the resonance voltages V_{LL} and V_{RR} immediately after the COOK synchronized single-cycle shorting. This is because the droop, caused by I^2R losses through the parasitic resistances R_2 and R_{sw} of L_2 and the data-modulation switch, decreases with Q_2. As a result, the Q_2 trade-off between PTE and data rate for LSK modulation is obviated for COOK modulation. As illustrated in Fig. 7.4(C), increasing Q_2 for greater PTE and energy savings reduces resonance recovery time and hence increases achievable data rate.

The initial increase of the recovery time for low Q_2 is due to longer time constants of the secondary resonant tank. At higher Q_2, the recovery time decreases even with longer time constant because of a reduced amplitude droop in the resonance voltages V_{LL} and V_{RR}. As Q_2 increases, the parasitic resistance R_2 of the inductor L_2 decreases and so does the amplitude droop. Thus, the recovery time decreases for higher Q_2.

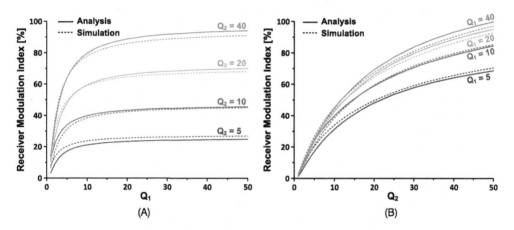

Figure 7.5 Analytical and simulated dependence of receiver modulation index of COOK modulation as a function of primary and secondary quality factors Q_1 and Q_2. The parameters are: $k = 0.15$, $L_1 = 3\ \mu H$, $L_2 = 470$ nH, $f_{res} = 13.56$ MHz, $C_1 = 46$ pF, $C_2 = 293$ pF, $R_{sw} = 1\ \Omega$, $R_{L,ac} = 1\ k\Omega$.

Fig. 7.5 shows the effect of Q_1 and Q_2 on receiver modulation index (RMI), expressed as

$$\text{RMI} = \Delta V_{L1}/V_{L1,A} \tag{7.1}$$

where ΔV_{L1} is the data-driven voltage increase in V_{L1}, and $V_{L1,A}$ is the baseline amplitude of V_{L1}, denoted in Fig. 7.3(C). RMI directly quantifies the effect of COOK data modulation on resolvable voltage differences at the receiver, and hence gives a measure for how well the data decoding can be performed. A detailed analysis of RMI is provided in the Appendix, the results of which are superimposed on the simulated data in Fig. 7.5. RMI increases both with increasing Q_1 and Q_2, although the dependence on Q_2 is much stronger.

In summary, maximizing quality factors of the inductive link in COOK modulation yields not only optimal PTE, but also maximizes data rate, maximizes decoding quality, and minimizes energy losses.

The principle of synchronous resonant data transmission also extends to the primary side for greater energy efficiency and data bandwidth of forward power and data telemetry. However, the series LC tank on the power transmitting side requires synchronous opening of the series tank [53] at the expense of stringent timing control at high-voltage power.

7.3.3 Symbol data encoding

As outlined in Sect. 7.3.1 and illustrated in Fig. 7.3(C), the COOK synchronized shorting of the secondary coil produces an immediate step increase ΔV_{L1} in peak voltage amplitude at the primary that can be detected for data decoding right upon completion of the cycle. Hence the timing of shorting cycles can be chosen to encode data symbols. For concurrent power delivery through full-wave rectification, the average number of non-shorting cycles in the encoding scheme must be maximized. Here we consider 4-cycle encoding schemes which ensure at least 6 non-shorting rectification half cycles out of 8 total half cycles, guaranteeing at least 75% of peak power delivery regardless of data. Simple binary encoding, shown in Fig. 7.6(A), permits transmission of a single bit for every four carrier cycles, at an average 7 non-shorting out of 8 half-cycles

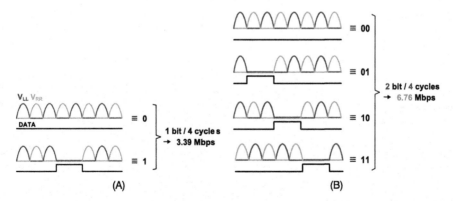

Figure 7.6 (A) Simple COOK modulation scheme with binary symbol data encoding. (B) Enhanced COOK modulation scheme with quaternary symbol data encoding resulting into two-fold increase in data rate. Both schemes support at least 6 out of 8 half cycles of power rectification over the same inductive link.

(87.5% rectification duty cycle). In contrast, varying the position of the single short-ing cycle within the four-cycle time interval allows enhanced data rate. As illustrated in Fig. 7.6(B), encoding of three positions at integer cycle offsets achieves two bits for every four carrier cycles, or a 6.78 Mbps data rate at half the 13.56-MHz carrier fre-quency, and an average of 6.5 non-shorting out of 8 half-cycles (81.25% rectification duty cycle).

7.3.4 Resonance recovery and data rate

Considerations of resonance recovery also govern the achievable data rate: energy in the tank needs to be sufficiently replenished between consecutive coil shortings. Hence the maximum data rate is inversely proportional to resonance recovery time. For the COOK modulation data encoding scheme at half the carrier frequency, the average interval between consecutive shortings is 4.33 carrier cycles (13 non-shorting cycles for every 3 shorting cycles). For this average interval, Fig. 7.4(C) shows that, regardless of the quality factor Q_2, the secondary tank recovers to greater than 95% of its original voltage amplitude between consecutive shortings.

7.4. System implementation

For the experimental validation a prototype COOK modulation IC and board-level telemetry system were implemented as shown in Fig. 7.7. The IC integrates a full-wave rectifier, clock recovery comparator, phase-locked loop (PLL), bias blocks, switch for data modulation, and auxiliary circuitry for data transmission and system control, as highlighted in the dashed box in Fig. 7.7. Outside the IC, the secondary side includes a parallel LC tank (L_2 and C_2), a load capacitor C_L, and a load resistor R_L for modeling load current. On the primary side, a series LC tank (L_1 and C_1) is located concentric to the parallel LC tank of the secondary side for power and data transfer. Measured

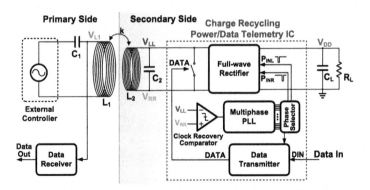

Figure 7.7 Block diagram of the implemented system for power and data telemetry.

geometries and parameters of the primary and secondary coils are as follows: $L_1 = 5.27\ \mu H$, $R_1 = 5.92\ \Omega$, $Q_1 = 75.85$, $D_1 = 6.5$ cm, $N_1 = 6$, $L_2 = 2.47\ \mu H$, $R_2 = 2.79\ \Omega$, $Q_2 = 75.43$, $D_2 = 4.2$ cm, and $N_2 = 8$ where $L_{1,2}$ represent the inductances, $R_{1,2}$ the parasitic resistances, $Q_{1,2}$ the quality factors, $D_{1,2}$ the diameters, and $N_{1,2}$ the numbers of windings for the primary and secondary coils, respectively, for each link. Values of the external surface-mount device (SMD) chip capacitors (C_1 and C_2) were chosen for 13.56-MHz resonance on both sides. For precise control of distance between the coils, plastic spacers of calibrated lengths were inserted in-between. Data bit streams were generated in field-programmable gate-array (FPGA) on the secondary-side board, and were fed to the telemetry IC capturing the data bit stream and producing the data pulse (DATA) to drive the modulation switch across the coil accordingly. On the primary side, a signal generator generated a 13.56-MHz sine wave input to the primary LC tank. The voltage V_{L1} across the primary coil was sampled with an oscilloscope and decoded in Matlab to determine bit error rates.

Detailed circuit diagrams of major system blocks are depicted in Figs. 7.8 and 7.10. As shown on the left side in Fig. 7.8(A), the common-gate comparator detects the timing when the voltage across the tank is zero for generation of synchronized single-cycle shortings by comparing the two tank voltages V_{LL} and V_{RR}. While comparators in conventional rectifiers compare coil voltages with V_{DD} to directly generate switch

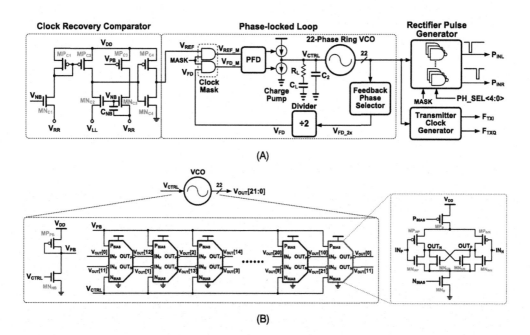

Figure 7.8 Circuit diagrams of (A) the comparator, PLL and pulse and clock generation blocks and (B) voltage-controlled oscillator (VCO).

signals for rectification, the comparator in this work is used for clock recovery as reference to a PLL generating the switching signals instead, lowering the comparator power consumption and design complexity.

The recovered clock from the comparator V_{REF} is fed as the reference clock to the 22-phase frequency-doubling type-2 PLL shown in Fig. 7.8(A). Its voltage-controlled oscillator (VCO) consists of 11 delay cell stages as depicted in Fig. 7.8(B). Currents to the delay cells are controlled through both pMOS and nMOS current sources for balancing the voltage range of the delay cell outputs. For differential operation, cross-coupled nMOS differential pairs are inserted across the differential outputs in each cell.

The PLL directly controls the timing of the data pulse (DATA) shorting the LC tank for data transmission. Hence PLL timing accuracy is critically important for maintaining low BER and high energy efficiency in data transmission. At the same time, the power consumption of the PLL should be contained to minimize the total power for overall energy efficiency. Simulations of the trade-off between LC tank power loss and VCO power consumption as a function of RMS VCO jitter in Fig. 7.9 demonstrate minimal impact on data transmission efficacy and overall energy efficiency for VCO jitter less than 1 ns. The measured jitter of 440 ps (Fig. 7.15(B)) is near the optimal trade-off point in Fig. 7.9.

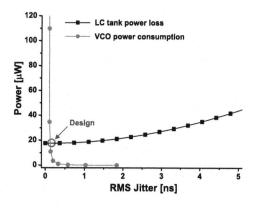

Figure 7.9 LC tank power loss and VCO power consumption as a function of the PLL RMS period jitter. The LC link was simulated in Cadence to obtain power loss of the secondary LC tank by varying the jitter. The PLL jitter was simulated using a behavioral simulator (CppSim [56]).

The PLL and the rectifier pulse generator produce the phase-tuned pulses P_{INL} and P_{INR} gating the pMOS switches of the full-wave rectifier shown in Fig. 7.10(A). During data transmission, the clocks to the PLL phase-frequency detector (PFD) are blocked by a mask signal not to disturb the PLL locking.

As illustrated in Fig. 7.11, the 22 phases of the VCO are aligned over each half-cycle of the resonant tank. Among the 22 phases, one is selected to generate P_{INL} and P_{INR} for full-wave rectification, alternating between V_{LL}-active and V_{RR}-active half-cycles.

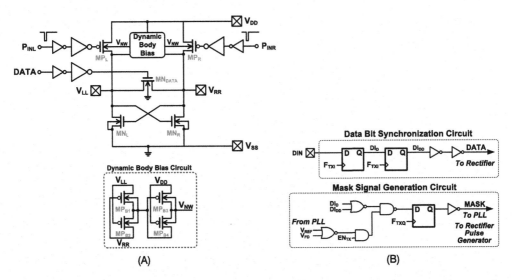

Figure 7.10 Circuit diagrams of (A) full-wave rectifier with data modulation switch across the LC tank, and (B) data bit synchronization and PLL/rectifier mask generation timing circuits.

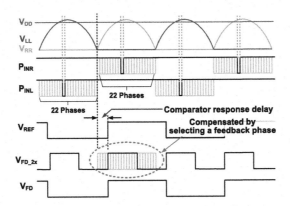

Figure 7.11 Timing diagram of the multiple phases of the PLL aligning with the LC tank signals V_{LL} and V_{RR}, the switching signals for rectifier P_{INL} and P_{INR}, and the PLL input and feedback signals V_{REF}, V_{FD_2x} and V_{FD}. Comparison delay of the clock recovery comparator shown in V_{REF} is compensated by feeding a delayed phase back to the PLL look.

Hence, any delay in the clock recovery due to the comparator can be compensated by selecting a shifted phase signal in the PLL feedback V_{FD_2x} to align the PFD feedback V_{FD} to the delayed recovered clock V_{REF}. As a result, design requirements on comparator delay can be relaxed to minimize power consumption.

The schematic of the full-wave rectifier with two cross coupled nMOS transistors and two pMOS switches driven by buffered phase pulses P_{INL} and P_{INR} is shown in Fig. 7.10(A). An additional nMOS switch across V_{LL} and V_{RR} is provided for COOK synchronous shorting by the buffered DATA signal during data transmission. While the body terminals of the nMOS transistors connect to the grounded substrate, the body terminals of the pMOS transistors share an *n*-well connected to a three-way dynamic body bias generator shown in the inset of Fig. 7.10(A). The dynamic body bias generator tracks the highest voltage among the three source/drain voltages: V_{LL}, V_{RR}, and V_{DD}.

As shown in Fig. 7.10(B), the data bit synchronization circuit receives external data bits (DIN) to generate the shorting signal (DATA), which drives the switch across the LC tank. As depicted in Fig. 7.12, this circuit synchronizes the data signal with F_{TXI}, generated from the PLL and transmitter clock generator as shown in Fig. 7.8(A).

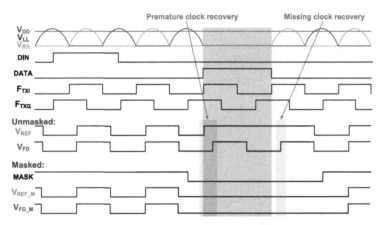

Figure 7.12 Timing diagram of data bit reception and synchronization (DIN and DATA), and mask signal generation for stable PLL and rectifier operation. By masking V_{REF} and V_{FD} during data bit transmission, clock edge misalignment and edge missing issues are resolved in the masked signals V_{REF_M} and V_{FD_M}.

Shorting of the LC tank signals (V_{LL} and V_{RR}) for data transmission may lead to missed clock recovery, the disturbance of the PLL loop, and current leakage in the rectifier. These potential problems are avoided in this design by generating a mask signal MASK in the circuit shown on the bottom of Fig. 7.10(B). As shown in Fig. 7.12, the recovered clock from the comparator serving as the PLL reference clock V_{REF} rises too early during the short, and misses the next rising edge after the short. By applying the MASK signal, V_{REF_M} and V_{FD_M} are free of false and missing clock edges into the PLL. Similarly, the mask signal is also applied to the rectifier to prevent any reverse current leakage during the shorts.

7.5. Measurement results

The power and data telemetry IC with the proposed COOK modulation scheme was fabricated in 65 nm CMOS. A micrograph of the IC is shown in Fig. 7.13. The rectifier

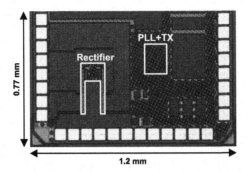

Figure 7.13 Chip micrograph.

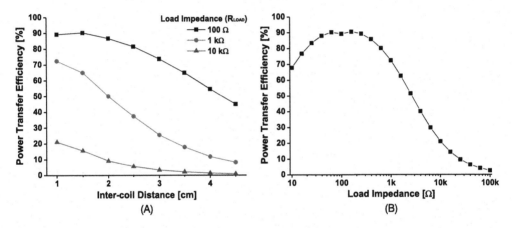

Figure 7.14 Measured power transfer efficiencies of the LC link. Geometries and parameters of the primary and secondary coils are as follows: $L_1 = 5.27$ μH, $R_1 = 5.92$ Ω, $Q_1 = 75.85$, $D_1 = 6.5$ cm, $N_1 = 6$, $L_2 = 2.47$ μH, $R_2 = 2.79$ Ω, $Q_2 = 75.43$, $D_2 = 4.2$ cm, and $N_2 = 8$. Values of the external surface-mount device (SMD) chip capacitors (C_1 and C_2) were chosen for 13.56-MHz resonance on both sides. R_{LOAD} is inserted across the secondary parallel LC tank to model the resistive load on the secondary side. (A) PTE as a function of distance between the primary and secondary coils at R_{LOAD} of 100 Ω, 1 kΩ and 10 kΩ. (B) PTE as a function of R_{LOAD} at 1-cm distance.

occupies an active area of 0.017 mm^2, and other circuits including the PLL, transmitter, and bias circuits measure 0.029 mm^2.

Fig. 7.14 validates the PTE of the inductive link used in this work, measured with the method outlined in [54,55] using a network analyzer. Note that the measured quality factors of both the primary and secondary inductors are higher than 75. At 100-Ω load impedance and 1-cm inter-coil distance, the measured PTE peaks at 89.2% and decreases with increasing distance as given in Fig. 7.14(A). Fig. 7.14(B) shows that the inductive link is optimized around 100-Ω load impedance, for about 10-mW power transfer at 1.2-V power supply.

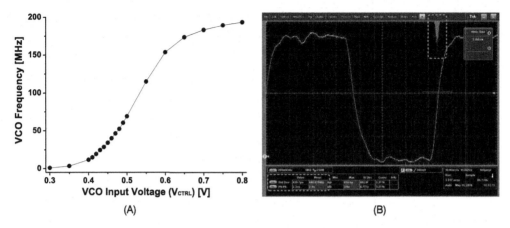

Figure 7.15 (A) Measured VCO frequency as a function of VCO input voltage. (B) RMS and peak-to-peak jitter of the two-fold frequency divided output signal of the PLL.

Fig. 7.15(A) shows measured VCO frequency over VCO input voltage V_{CTRL}. At a V_{CTRL} around 0.45 V, the PLL output frequency is at its target of 27.12 MHz. The corresponding PLL phase noise is 104.1 dBc at 1 MHz offset. Fig. 7.15(B) shows the measurement and histogram of the period jitter in F_{TXI}, the two-fold frequency divided signal output of the PLL. This signal at 13.56 MHz is directly utilized to generate the modulation pulse. Measured RMS jitter is 440 ps and peak-to-peak jitter is 2.3 ns, which contributes less than 4% of overall LC tank energy losses based on the results shown in Fig. 7.9. There the measured jitter is well within acceptable levels such that it does not contribute to reduction in energy efficiency and timing integrity in data transmission.

Fig. 7.16 shows typical recorded secondary and primary coil voltage waveforms during data transmission. As shown in the figure, the IC shorts the LC tank at times when the LC tank voltages V_{LL} and V_{RR} coincide, and maintains the shorts for single cycles. As expected, because the initial phase of the LC resonant oscillation is resumed at the end of each shorting, the LC oscillation recovers to the original amplitude within a few cycles. In the bottom of Fig. 7.16, data-driven responses induced by the single-cycle shortings on the primary side are also shown. At V_{L1} the voltage across the primary coil L_1, step increases of the peak voltages are induced by the shortings on the secondary side. These step increases are clearly distinguishable in the amplitude and timing of the voltage waveform. Other than these cycles with data-driven step increases, the peak amplitude of V_{L1} decreases monotonically or reaches an asymptote. Hence, reliable data decoding can be performed by detecting step increases in the peak voltages of V_{L1} beyond a positive threshold.

As a proof-of-concept for data transmission, a pseudo-random data bit stream was generated from sequence combinations of 16 2-bit data units, where the sequence in-

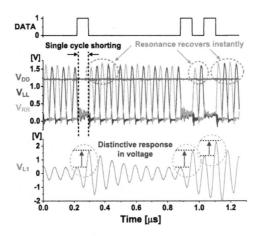

Figure 7.16 Measured voltage waveforms of COOK modulation scheme. The secondary LC tank voltages V_{LL} and V_{RR} recover their amplitudes right after data-driven single-cycle shortings. In response to the shortings, clear voltage increases are induced at V_{L1} on the primary side.

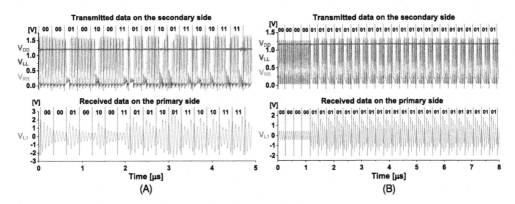

Figure 7.17 Measured voltage waveforms for data transmission and reception at a rate of 6.78 Mbps with (A) a pseudo-random bit sequence, and (B) a repeating "01" pattern sequence.

cluded all possible transitions in the COOK 2-bit encoded modulation, i.e., 00–00, 00–01, 00–10, . . . , 11–10, and 11–11. Fig. 7.17(A) shows measured waveforms on the secondary and primary sides during transmission of pseudo–random data bit stream at a data rate of 6.78 Mbps, at half the carrier frequency of 13.56 MHz. The top panel of Fig. 7.17(A) shows voltage waveforms V_{LL}, V_{RR} and V_{DD} on the secondary side during data transmission, showing pulse position encoding of the data over a four-cycle window. As expected, the resulting step increases in the peak of voltage V_{L1} at the primary in the bottom panel of Fig. 7.17(A) align with the timing of shortings on the transmitter side. Fig. 7.17(B) further validates that the data transmission is robust to effects of recurring patterns in the bit stream such as repeating sequences of "01" in the data.

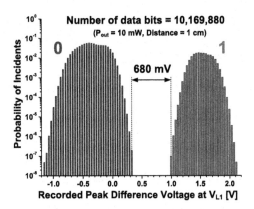

Figure 7.18 Histogram of the peak value differences of V_{L1} at $P_{out} = 10$ mW and coil distance $d = 1$ cm. More than 10 million bits were transmitted and no error was detected. "1" represents the peak voltage difference when shorted, and "0" when not shorted. A large voltage gap of 680 mV is observed between the two groups.

For extensive validation, over several million bits were transmitted over the COOK link and bit-error rates (BERs) were measured under varying load, distance, and alignment conditions. Fig. 7.18 shows recorded peak-difference in voltage V_{L1} at 1 cm distance d between the two coils and at 10 mW of output power P_{out} simultaneously delivered to the load on the secondary side. Under these conditions, over 10 million bits were transferred without error in decoding, for a BER $\leq 9.84 \times 10^{-8}$. As shown in the figure, the voltage gap between the two groups of detected data "0" and "1" is 680 mV, providing large margin for threshold decoding. Accounting for this margin, the achievable BER by threshold decoding estimated from a Gaussian fit of the empirical probability distribution shown in Fig. 7.18 is 2.3×10^{-11}.

The experiment was repeated, each over 1 million transmitted bits at the same 6.78-Mbps data rate, varying three parameters: output power delivered to the load P_{out}, inter-coil distance d, and lateral misalignment between the two coils. Fig. 7.19 shows the BER and RMI varying P_{out} from 100 μW to 11.5 mW at fixed 1-cm inter-coil distance. At each delivered power level, again no errors were detected. The estimated BERs using the same Gaussian-fit procedure are given in Fig. 7.19. As expected, the BER increases and RMI decreases as P_{out} increases. Up to 11.5 mW of delivered power, a BER better than 10^{-6} can be achieved.

Fig. 7.20 shows BER and RMI for varying distance d between the coils, at fixed P_{out} of 1 mW. Up to 3.5 cm distance, again no errors were detected with more than 1 million bits transmitted, validating robust operation of data and power telemetry for the given coil geometries. Gaussian-estimated BERs from the voltage measurements are shown for these distances up to 3.5 cm, whereas the empirical BER measures are shown for distances from 4 to 5 cm.

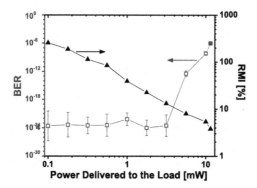

Figure 7.19 Bit-error rates (BER) and receiver modulation indices (RMI) as a function of P_{out}, the power delivered to the load, at coil distance $d = 1$ cm. In these and the following BER experiments, filled box symbols indicate direct measurement of BER as the frequency of errors, while empty box symbols indicate estimated BER in the absence of observed errors. For the BER estimates, mean and standard deviation are shown as obtained from Gaussian fit extrapolation of the "0" and "1" histograms in Fig. 7.18.

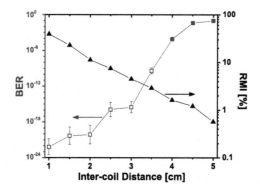

Figure 7.20 Bit-error rates (BER) and receiver modulation indices (RMI) as a function of coil distance at $P_{out} = 1$ mW.

Fig. 7.21 shows BER and RMI for varying lateral misalignment in the axial distance between the two coils at fixed $d = 1$ cm and $P_{out} = 1$ mW, again validating robust error-free operation up to 2.5 cm in lateral misalignment, which is 77% of the primary coil radius.

Fig. 7.22 illustrates the effect of data transmission on power delivery, revealing a perhaps counterintuitive benefit. Shown is the ratio of the amount of power delivered to the load in both cases: P_{out}^{data} during data transmission ("on") and $P_{out}^{no\ data}$ without data transmission ("off"). During data transmission, on average three shortings of the LC tank occur every 16 cycles, so power is delivered only a 81.25% fraction of the time, and a corresponding reduction in the power ratio $P_{out}^{data}/P_{out}^{no\ data}$ may be expected between the "on" and "off" conditions. However, a P_{out}^{data} greater than 81.25% of $P_{out}^{no\ data}$

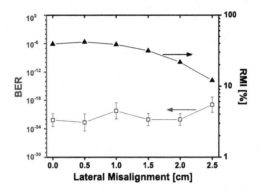

Figure 7.21 Bit-error rates (BER) and receiver modulation indices (RMI) as a function of lateral coil alignment at $P_{out} = 1$ mW and $d = 1$ cm.

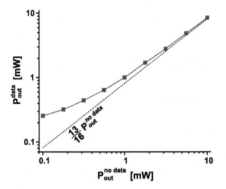

Figure 7.22 Ratio of $P_{out}^{no\,data}$ (the delivered power to the load with data transmission *off*) and P_{out}^{data} (the delivered power to the load with data transmission *on*). The black dashed line indicates $P_{out}^{no\,data}$ multiplied by 81.25% (13/16), the average duty cycle of non-shorting cycles for data transmission at half-carrier 6.78 Mpbs data rate. As a side effect of COOK modulation, transients across the resonant tank cause the primary to deliver more power to the secondary, compensating for off-time in rectification during modulation.

is observed at all measured power levels, and even greater than $P_{out}^{no\,data}$ at power levels less than 1 mW. The cause for the power boost in the "on" condition can be inferred from Figs. 7.16 and 7.17. As shown, each cycle shorting induces increased voltage and current swing on the primary side, which persist over several cycles, and which through induction increase the voltage and current in the secondary coil as well. The result is that during non-shorting cycles actually more power is delivered from the primary to the secondary, amply compensating for the skipped power delivery during shorting cycles. At smaller levels of $P_{out}^{no\,data}$, the RMI increases with a larger voltage increase in the primary, resulting in increased power delivery P_{out}^{data} to the secondary.

Table 7.1 Performance comparison of inductive uplink telemetry systems.

Reference	Number of links	Modulation scheme	Carrier freq. [MHz]	Data rate [Mbps]	Carrier cycle per bit	Bit-error rate	Energy per bit [pJ/bit]
[49]	Single	LSK	13.56	0.1	135.6	N/A	N/A
[50]		LSK	13.56	0.5	27.1	N/A	N/A
[51]		PPSK	13.56	0.85	16	N/A	N/A
[12]		LSK	400	1	400	$< 1.7 \times 10^{-7}$	N/A
[39]	Multiple	LSK	25	2.8	8.9	$< 10^{-6}$	35.7
[27]		BPSK	48	3	16	2×10^{-4}	1962
This work	Single	COOK	13.56	6.78	2	$< 9.9 \times 10^{-8}$	9.5

This phenomenon can be partially interpreted as a form of energy recycling. A single cycle shorting delays transmission of energy from the primary for a few cycles, which circulates in the primary and is then added to the energy transmitted in subsequent cycles. The increase in P_{out}^{data} relative to $P_{out}^{no\ data}$ owes to a larger fraction of the secondary tank power captured by the rectifier during transient higher voltage levels.

Table 7.1 compares the COOK modulation scheme with state-of-the-art uplink telemetry. As expected, conventional single-link modulation schemes achieve relatively low data rates. Multiple-link schemes are capable of achieving a few Mbps with relatively low BER ($< 10^{-6}$ for [39] and 2×10^{-4} for [27]), at the expense of the need for a separate link for power transfer. In contrast, single-link COOK excels in data rate even compared to the multiple-link schemes, achieving a data rate at half the carrier frequency with one of the lowest BERs. The high Q of the inductive link permits 11.5-mW power delivery to the load simultaneously with 6.78-Mbps data over 1-cm distance. The measured total power consumption of the circuits including the PLL, transmitter, bias, and switch driving buffers is 64.44 µW at 6.78-Mbps data rate, resulting in an energy per transmitted bit, as total consumed energy divided by data rate, of 9.50 pJ/bit, an order of magnitude lower than the state-of-the-art.

7.6. Conclusion

We have presented and demonstrated cyclic on-off keying (COOK) as a new modulation scheme for simultaneous transmission of power and broadband data over the same resonant inductive link. The key principle of COOK is conservation of energy in data-synchronous single-cycle adiabatic switching of the LC resonant tank during data transmission, minimizing power losses while also minimally disturbing LC resonance conditions. Time-based encoding of the data allows to transmit two bits per four carrier cycles, substantially larger than conventional transmission data encoding schemes that deplete and hence require recovery of LC tank energy.

7.7. Appendix

In this section we derive the receiver modulation index RMI for COOK modulation. As explained in Sect. 7.3.2 and expressed in Eq. (7.1), RMI quantifies the profile of amplitude change in the primary coil voltage V_{L1} during the synchronized COOK shorting for data transmission. Since there are two distinctive causes for the amplitude change, RMI can be expressed as

$$\text{RMI} = \text{RMI}_Z + \text{RMI}_{exp} \tag{7.2}$$

where RMI_Z represents amplitude change due to reflected impedance change by single-cycle shorting, and RMI_{exp} the change in V_{L1} incurred by the exponentially decreasing current through the secondary inductor during the shorting.

The increased portion of the amplitude due to change of the reflected impedance Z_{rfl} in RMI_Z is reflected from the secondary side on to the primary side. The input impedance Z_{in} from the voltage source is expressed as

$$Z_{in} = Z_1 + Z_{rfl} = Z_1 - \frac{s^2 M^2}{Z_2(s)}. \tag{7.3}$$

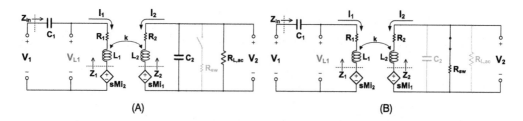

Figure 7.23 Analytical circuit models when the data switch is (A) open and (B) shorted.

At the resonance frequency, and while the switch is open as shown in Fig. 7.23(A), Z_{in} can be expressed as

$$
\begin{aligned}
Z_{in}^o &= \frac{1}{j\omega C_1} + j\omega L_1 + R_1 + Z_{rfl}^o \\
&\simeq R_1 \left(1 + k^2 Q_1 R_1 Q_{2L} \right)
\end{aligned} \tag{7.4}
$$

where $Q_{2L} = \omega L_2 / (R_2 + \frac{L_2}{C_2 R_{L,ac}})$.

Conversely, Z_{rfl} and Z_{in} during the short as shown in Fig. 7.23(B) can be expressed as

$$Z_{rfl}^s = \frac{k^2 \omega^2 L_1 L_2 (R_2 + R_{sw})}{(R_2 + R_{sw})^2 + (\omega L_2)^2} + j\omega L_1 \frac{k^2 \omega^2 L_2^2}{(R_2 + R_{sw})^2 + (\omega L_2)^2}, \tag{7.5}$$

$$Z_{in}^{s} = \frac{1}{j\omega C_1} + j\omega L_1 \left(1 - k^2 \frac{Q_{2sw}^2}{Q_{2sw}^2 + 1}\right) + R_1 \left(1 + k^2 \frac{Q_1 Q_{2sw}}{Q_{2sw}^2 + 1}\right) \tag{7.6}$$

where $Q_{2sw} = \omega L_2 / (R_2 + R_{sw})$.

Due to this impedance change from Z_{in}^o to Z_{in}^s, the voltage amplitude is increasing over the single cycle shorting as follows:

$$\begin{aligned} \mathrm{RMI}_Z &= \left(\frac{1 + Z_{in}^o}{1 + Z_{in}^s} - 1\right)\left(1 - e^{-T\frac{R_{1s}}{2L_{1s}}}\right) \\ &= \left(\frac{Z_{in}^o - Z_{in}^s}{1 + Z_{in}^s}\right)\left(1 - e^{-\pi/Q_{1s}}\right) \end{aligned} \tag{7.7}$$

where T is the resonance period, and where

$$Q_{1s} = \omega L_{1s}/R_{1s} = \frac{\omega L_1 \left(1 - k^2 \frac{Q_{2sw}^2}{Q_{2sw}^2 + 1}\right)}{R_1 \left(1 + k^2 \frac{Q_1 Q_{2sw}}{Q_{2sw}^2 + 1}\right)}. \tag{7.8}$$

RMI_{exp} is induced by the current I_2 flowing through L_2, R_2 and the switch (R_{sw}) during the single-cycle shorting. At the start of the short, the current through the secondary inductor is at its maximum I_{2max}. During the short, the current decreases exponentially with time constant formed by L_2, R_2 and R_{sw} as follows:

$$I_2(t) = I_{2max} e^{-t\left(\frac{R_2 + R_{sw}}{L_2}\right)}. \tag{7.9}$$

This exponential decaying current is reflected to V_{L1} during the single cycle shorting as follows:

$$\mathrm{RMI}_{exp} = sMI_2(T)/V_{L1}^o = -\frac{\eta M e^{-2\pi\left(\frac{1}{Q_2} + \frac{1}{Q_{2sw}}\right)}}{L_1 + \eta M}, \tag{7.10}$$

with current peak conversion ratio $\eta = I_{2max} / I_{1max} = k\sqrt{R_1 Q_1 Q_{2L} / (R_2 + \frac{L_2}{C_2 R_{L,ac}})}$ and link mutual inductance $M = k\sqrt{L_1 L_2}$.

References

[1] M.W. Baker, R. Sarpeshkar, Feedback analysis and design of RF power links for low-power bionic systems, IEEE Transactions on Biomedical Circuits and Systems 1 (1) (2007) 28–38.

[2] U.-M. Jow, M. Ghovanloo, Design and optimization of printed spiral coils for efficient transcutaneous inductive power transmission, IEEE Transactions on Biomedical Circuits and Systems 1 (3) (2007) 193–202.

[3] A.K. RamRakhyani, S. Mirabbasi, C. Mu, Design and optimization of resonance-based efficient wireless power delivery systems for biomedical implants, IEEE Transactions on Biomedical Circuits and Systems 5 (1) (2011) 48–63.

[4] M. Zargham, P.G. Gulak, Maximum achievable efficiency in near-field coupled power-transfer systems, IEEE Transactions on Biomedical Circuits and Systems 6 (3) (2012) 228–245.

[5] S.Y. Hui, Planar wireless charging technology for portable electronic products and Qi, Proceedings of the IEEE 101 (6) (2013) 1290–1301.

[6] R.F. Xue, K.W. Cheng, M. Je, High-efficiency wireless power transfer for biomedical implants by optimal resonant load transformation, IEEE Transactions on Circuits and Systems I: Regular Papers 60 (4) (2013) 867–874.

[7] I. Mayordomo, T. Drager, P. Spies, J. Bernhard, A. Pflaum, An overview of technical challenges and advances of inductive wireless power transmission, Proceedings of the IEEE 101 (6) (2013) 1302–1311.

[8] H. Xu, J. Handwerker, M. Ortmanns, Telemetry for implantable medical devices: Part 2 – power telemetry, IEEE Solid-State Circuits Magazine 6 (3) (2014) 60–63.

[9] B.S. Wilson, M.F. Dorman, Cochlear implants: a remarkable past and a brilliant future, Hearing Research 242 (1–2) (2008) 3–21.

[10] L. Yi-Kai, C. Kuanfu, P. Gad, L. Wentai, A fully-integrated high-compliance voltage SoC for epi-retinal and neural prostheses, IEEE Transactions on Biomedical Circuits and Systems 7 (6) (2013) 761–772.

[11] H. Bhamra, Y. Kim, J. Joseph, J. Lynch, O.Z. Gall, H. Mei, C. Meng, J. Tsai, P. Irazoqui, A 24 μW batteryless, crystal-free, multinode synchronized SoC "Bionode" for wireless prosthesis control, IEEE Journal of Solid-State Circuits 50 (11) (2015) 2714–2727.

[12] R. Muller, H.-P. Le, W. Li, P. Ledochowitsch, S. Gambini, T. Bjorninen, A. Koralek, J.M. Carmena, M.M. Maharbiz, E. Alon, J.M. Rabaey, A minimally invasive 64-channel wireless μECoG implant, IEEE Journal of Solid-State Circuits 50 (1) (2015) 344–359.

[13] A. Yakovlev, J.H. Jang, D. Pivonka, An 11 μW sub-pJ/bit reconfigurable transceiver for mm-sized wireless implants, IEEE Transactions on Biomedical Circuits and Systems 10 (1) (2016) 175–185.

[14] Y.P. Lin, C.Y. Yeh, P.Y. Huang, Z.Y. Wang, H.H. Cheng, Y.T. Li, C.F. Chuang, P.C. Huang, K.T. Tang, H.P. Ma, Y.C. Chang, S.R. Yeh, H. Chen, A battery-less, implantable neuro-electronic interface for studying the mechanisms of deep brain stimulation in rat models, IEEE Transactions on Biomedical Circuits and Systems 10 (1) (2016) 98–112.

[15] Z. Xiao, X. Tan, X. Chen, S. Chen, Z. Zhang, H. Zhang, J. Wang, Y. Huang, P. Zhang, L. Zheng, H. Min, An implantable RFID sensor tag toward continuous glucose monitoring, IEEE Journal of Biomedical and Health Informatics 19 (3) (2015) 910–919.

[16] W.B. Spillman, Sensing and processing for smart structures, Proceedings of the IEEE 84 (1) (1996) 68–77.

[17] A. Abrial, J. Bouvier, M. Renaudin, P. Senn, P. Vivet, A new contactless smart card IC using an on-chip antenna and an asynchronous microcontroller, IEEE Journal of Solid-State Circuits 36 (7) (2001) 1101–1107.

[18] P. Merlino, A. Abramo, An integrated sensing/communication architecture for structural health monitoring, IEEE Sensors Journal 9 (11) (2009) 1397–1404.

[19] S. Jiang, S.V. Georgakopoulos, Optimum wireless power transmission through reinforced concrete structure, in: Proceedings of IEEE International Conference on RFID, 2011, pp. 50–56.

[20] D. Li, M. Shen, J. Huangfu, J. Long, Y. Tao, J. Wang, C. Li, L. Ran, Wireless sensing system-on-chip for near-field monitoring of analog and switch quantities, IEEE Transactions on Industrial Electronics 59 (2) (2012) 1288–1299.

[21] D. Cirmirakis, A. Demosthenous, N. Saeidi, N. Donaldson, Humidity-to-frequency sensor in CMOS technology with wireless readout, IEEE Sensors Journal 13 (3) (2013) 900–908.

[22] R. Bhattacharyya, C. Floerkemeier, S. Sarma, Low-cost, ubiquitous RFID-tag-antenna-based sensing, Proceedings of the IEEE 98 (9) (2010) 1593–1600.

[23] B.S. Cook, R. Vyas, K. Sangkil, T. Trang, L. Taoran, A. Traille, H. Aubert, M.M. Tentzeris, RFID-based sensors for zero-power autonomous wireless sensor networks, IEEE Sensors Journal 14 (8) (2014) 2419–2431.

[24] L.H. Jung, P. Byrnes-Preston, R. Hessler, T. Lehmann, G.J. Suaning, N.H. Lovell, A dual band wireless power and FSK data telemetry for biomedical implants, in: Proceedings of the Annual International Conference of the IEEE Engineering in Medicine and Biology Society, 2007, pp. 6596–6599.

[25] G. Simard, M. Sawan, D. Massicotte, High-speed OQPSK and efficient power transfer through inductive link for biomedical implants, IEEE Transactions on Biomedical Circuits and Systems 4 (3) (2010) 192–200.

[26] L. Zheng, K. Chen, W. Liu, A non-coherent versatile DPSK receiver for high channel-density neural prosthesis, in: Proceedings of the IEEE Custom Integrated Circuits Conference, 2011, pp. 1–4.

[27] A.D. Rush, P.R. Troyk, A power and data link for a wireless-implanted neural recording system, IEEE Transactions on Biomedical Engineering 59 (11) (2012) 3255–3262.

[28] M. Kiani, M. Ghovanloo, A 13.56-Mbps pulse delay modulation based transceiver for simultaneous near-field data and power transmission, IEEE Transactions on Biomedical Circuits and Systems 9 (1) (2015) 1–11.

[29] C.S. Mestais, G. Charvet, F. Sauter-Starace, M. Foerster, D. Ratel, A.L. Benabid, WIMAGINE: wireless 64-channel ECoG recording implant for long term clinical applications, IEEE Transactions on Neural Systems and Rehabilitation Engineering 23 (1) (2015) 10–21.

[30] A.M. Sodagar, K.D. Wise, K. Najafi, A wireless implantable microsystem for multichannel neural recording, IEEE Transactions on Microwave Theory and Techniques 57 (10) (2009) 2565–2573.

[31] K.M. Silay, C. Dehollain, M. Declercq, Inductive power link for a wireless cortical implant with two-body packaging, IEEE Sensors Journal 11 (11) (2011) 2825–2833.

[32] R.R. Harrison, P.T. Watkins, R.J. Kier, R.O. Lovejoy, D.J. Black, B. Greger, F. Solzbacher, A low-power integrated circuit for a wireless 100-electrode neural recording system, IEEE Journal of Solid-State Circuits 42 (1) (2007) 123–133.

[33] M.S. Chae, Y. Zhi, M.R. Yuce, L. Hoang, W. Liu, A 128-channel 6 mW wireless neural recording IC with spike feature extraction and UWB transmitter, IEEE Transactions on Neural Systems and Rehabilitation Engineering 17 (4) (2009) 312–321.

[34] A.M. Sodagar, G.E. Perlin, Y. Ying, K. Najafi, K.D. Wise, An implantable 64-channel wireless microsystem for single-unit neural recording, IEEE Journal of Solid-State Circuits 44 (9) (2009) 2591–2604.

[35] F. Shahrokhi, K. Abdelhalim, D. Serletis, P.L. Carlen, R. Genov, The 128-channel fully differential digital integrated neural recording and stimulation interface, IEEE Transactions on Biomedical Circuits and Systems 4 (3) (2010) 149–161.

[36] S. Ha, J. Park, Y.M. Chi, J. Viventi, J. Rogers, G. Cauwenberghs, 85 dB dynamic range 1.2 mW 156 kS/s biopotential recording IC for high-density ECoG flexible active electrode array, in: Proceedings of the European Solid-State Circuits Conference, 2013, pp. 141–144.

[37] A. Borna, K. Najafi, A low power light weight wireless multichannel microsystem for reliable neural recording, IEEE Journal of Solid-State Circuits 49 (2) (2014) 439–451.

[38] M. Ghovanloo, S. Atluri, A wide-band power-efficient inductive wireless link for implantable microelectronic devices using multiple carriers, IEEE Transactions on Circuits and Systems I: Regular Papers 54 (10) (2007) 2211–2221.

[39] S. Mandal, R. Sarpeshkar, Power-efficient impedance-modulation wireless data links for biomedical implants, IEEE Transactions on Biomedical Circuits and Systems 2 (4) (2008) 301–315.

[40] F. Inanlou, M. Kiani, M. Ghovanloo, A 10.2 Mbps pulse harmonic modulation based transceiver for implantable medical devices, IEEE Journal of Solid-State Circuits 46 (6) (2011) 1296–1306.

[41] M. Kiani, M. Ghovanloo, A 20-Mb/s pulse harmonic modulation transceiver for wideband near-field data transmission, IEEE Transactions on Circuits and Systems II: Express Briefs 60 (7) (2013) 382–386.

[42] G. Wang, P. Wang, Y. Tang, W. Liu, Analysis of dual band power and data telemetry for biomedical implants, IEEE Transactions on Biomedical Circuits and Systems 6 (3) (2012) 208–215.

[43] S. Gambini, J. Crossley, E. Alon, J.M. Rabaey, A fully integrated, 290 pJ/bit UWB dual-mode transceiver for cm-range wireless interconnects, IEEE Journal of Solid-State Circuits 47 (3) (2012) 586–598.

[44] P.P. Mercier, S. Bandyopadhyay, A.C. Lysaght, K.M. Stankovic, A.P. Chandrakasan, A sub-nW 2.4 GHz transmitter for low data-rate sensing applications, IEEE Journal of Solid-State Circuits 49 (7) (2014) 1463–1474.

[45] A. Ebrazeh, P. Mohseni, 30 pJ/b, 67 Mbps, centimeter-to-meter range data telemetry with an IR-UWB wireless link, IEEE Transactions on Biomedical Circuits and Systems 9 (3) (2015) 362–369.

[46] Z. Tang, B. Smith, J.H. Schild, P.H. Peckham, Data transmission from an implantable biotelemeter by load-shift keying using circuit configuration modulator, IEEE Transactions on Biomedical Engineering 42 (5) (1995) 524–528.

[47] M. Ghovanloo, S. Atluri, An integrated full-wave CMOS rectifier with built-in back telemetry for RFID and implantable biomedical applications, IEEE Transactions on Circuits and Systems I: Regular Papers 55 (10) (2008) 3328–3334.

[48] A. Djemouai, M. Sawan, Prosthetic Power Supplies, Vol. 17, Wiley, New York, 1999, pp. 413–421.

[49] W. Xu, Z. Luo, S. Sonkusale, Fully digital BPSK demodulator and multilevel LSK back telemetry for biomedical implant transceivers, IEEE Transactions on Circuits and Systems II: Express Briefs 56 (9) (2009) 714–718.

[50] H.-M. Lee, M. Ghovanloo, An integrated power-efficient active rectifier with offset-controlled high speed comparators for inductively powered applications, IEEE Transactions on Circuits and Systems I: Regular Papers 58 (8) (2011) 1749–1760.

[51] D. Cirmirakis, J. Dai, A. Demosthenous, N. Donaldson, T. Perkins, A fast passive phase shift keying modulator for inductively coupled implanted medical devices, in: Proceedings of the European Solid-State Circuits Conference, 2012, pp. 301–304.

[52] S. Ha, C. Kim, J. Park, S. Joshi, G. Cauwenberghs, Energy-recycling integrated 6.78-Mbps data 6.3-mW power telemetry over a single 13.56-MHz inductive link, in: Symposium on VLSI Circuits Digest of Technical Papers, 2014, pp. 66–67.

[53] P.R. Troyk, W. Heetderks, M. Schwan, G. Loeb, Suspended carrier modulation of high-Q transmitters, ÖÖ U.S. Patent 5 697 076, Dec. 9, 1997.

[54] M. Kiani, M. Ghovanloo, A figure-of-merit for designing high-performance inductive power transmission links, IEEE Transactions on Industrial Electronics 60 (11) (2013) 5292–5305.

[55] D. Ahn, M. Ghovanloo, Optimal design of wireless power transmission links for millimeter-sized biomedical implants, IEEE Transactions on Biomedical Circuits and Systems 10 (1) (2016) 125–137.

[56] CppSim system simulator, [Online]. Available: http://www.cppsim.com.

CHAPTER 8

Fully integrated modular ECoG recording and stimulation

Contents

8.1. State-of-the-art ECoG interface systems

Various implantable devices for ECoG interfaces have been developed for clinical use and neuroscience research. Their target applications include treatment of neurological disorders and ECoG-based BCIs. Fig. 8.1 illustrates several state-of-the-art ECoG interface systems for clinical and research applications.

On the clinical side, implantable devices shown in Fig. 8.1(A)–(C) have been developed mostly for use in closed-loop treatment of intractable epilepsy as an alternative to tissue resection. These devices monitor ECoG signals and deliver stimulation to the seizure foci in response to epileptic seizure detection. The NeuroPace® RNS® System, shown in Fig. 8.1(A), is the first such system to receive FDA approval for closed-loop treatment in epilepsy patients, proven effective in reducing the frequency of partial-onset seizures in human clinical trials [10,11].

However, clinically proven devices are severely limited in the number of ECoG channels, typically less than ten, and rely on batteries, limiting implantation lifetime up to a few years. In addition, their physical size is too large to be implanted near the brain, so the main parts of these systems are implanted under the chest with a wired connection to the brain. Further developments in ECoG technology have striven to conquer these challenges: increasing number of channels, wireless powering, and miniaturization.

One such device is the Wireless Implantable Multi-channel Acquisition system for Generic Interface with NEurons (WIMAGINE), which features up to 64 channels, targeting long-term ECoG recording fully implanted in human patients [25] as shown in Fig. 8.1(D). This active IMD (AIMD) is fully covered by a 50-mm diameter hermetic housing made of silicone-coated titanium with a silicone–platinum electrode array on the bottom side. Its silicone over-molding is extended to include two antennas for

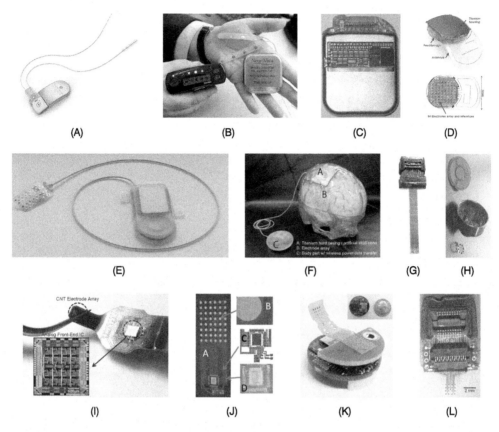

Figure 8.1 State-of-the-art ECoG interfacing systems. (A) NeuroPace RNS System [1,2]. (B) NeuroVista Seizure Advisory System [3–5]. (C) The neural interface (NI) system of Medtronic [6,7]. (D) The Wireless Implantable Multi-channel Acquisition system for Generic Interface with NEurons (WIMAGINE) [25]. (E) BrainCon system for a general-purpose medical BCI [12–14]. (F) The Wireless Human ECoG-based Real-time BMI System (W-HERBS) [8,9]. (G) µECoG recording system of Cortera Neurotechnologies, Inc. [15,22]. (H) An ECoG recording system with bidirectional capacitive data Telemetry [16]. (I) An ECoG recording system with a carbon nanotube microelectrode array and a corresponding IC [17]. (J) A wireless ECoG interface system with a 64-channel ECoG recording application-specific integrated circuit (ASIC) [18]. (K) An 8-channel low-cost wireless neural signal acquisition system made with off-the-shelf components [19]. (L) µECoG recording system with an electrode array fabricated on a transparent polymer for optogenetics-based stimulation [24].

RF communication and wireless power transfer. The housing fits inside of a 50-mm craniotomy, and its upper surface is just below the skin, with the implant replacing the previously existing bone. Two 32–channel ECoG recording ASICs [20] are implemented for ECoG recording, and commercial off-the-shelf components are employed for data processing, communication, power management, etc., leading to relatively high power consumption – 75 mW for 32 channels. Its operation and biocompatibility has been evaluated *in vivo* in non-human primates.

Another example is the BrainCon system with 16-channel ECoG recording and 8-channel stimulation designed for chronic implanted use in closed-loop human BCI [14]. Shown in Fig. 8.1(E), this device consists of an ECoG electrode array and an electronic package with a magnet, an inductive coil, and electronic components for data acquisition, stimulation and communication. Targeted for long-term recording and cortical stimulation in human patients, it is enclosed in a hermetic package with medical grade silicone rubber [12,13], and was validated *in vivo* for more than 10 months [14].

Further miniaturization and advances in functionality have been pursued through integration of circuits for ECoG recording, wireless powering, and wireless communication as shown in Fig. 8.1(F)–(J). In addition, low-cost ECoG interfaces shown in Fig. 8.1(K) for acute animal research have been developed [19], as have ECoG interfaces with transparent electrode arrays for compatibility with optogenetic stimulation shown in Fig. 8.1(L) [24].

Each of these devices offers substantial advances in wireless and integrated ECoG technology with improved functionality and increased density and channel counts. Yet, most rely on substantial cabling in connecting to the array of electrodes, or at least a wired connection to a distal ground as reference.

8.2. Encapsulated neural interfacing acquisition chip (ENIAC)

As highlighted in the previous section, most current state-of-the-art ECoG ASICs rely on external components, such as flexible substrates, electrode arrays, and antennas [21, 13,14,16,18,22–25]. In doing so, integration of all the components into a complete system requires special fabrication processes, and, importantly, requires a large number of connections between the readout ASIC and the electrodes, which are very difficult to manage in a hermetic environment. Furthermore, electrodes typically make direct metal–electrolyte contact to the surrounding tissue, which can lead to generation of toxic byproducts during electrical stimulation. In addition, most systems do not support electrical stimulation, while those that do offer limited stimulation efficiency, or require large external components for efficient operation. Finally, the silicon area occupied by the ASIC limits the span and density of electrodes across the cortical surface. For ultra-high channel count experiments, as needed for next-generation neuroscience and called for by several brain initiatives, such limitations must be overcome.

Instead of separating the electrodes and the ASIC, a promising approach that we present below is to integrate everything on a single encapsulated neural interfacing and acquisition chip (ENIAC), including electrodes, antennas for power and data telemetry, and all other circuits and components [26]. Thus, no external wires, substrates, batteries, or any other external components are required. Complete encapsulation of the ENIAC with a biocompatible material removes direct contact to tissue, including the electrodes for recording and stimulation. As such, the chip itself serves as a complete stand-alone neural interfacing system.

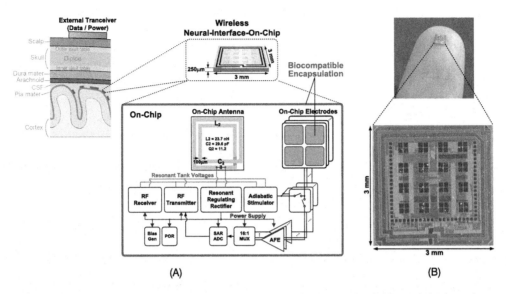

Figure 8.2 Encapsulated neural interfacing acquisition chip (ENIAC) [26,27]. (A) System diagram showing the fully integrated functionality of the ENIAC comprising on-chip antenna, electrodes, and all the circuitry for power management, communication, ECoG recording and stimulation. No external components are needed, and galvanic contact to surrounding tissue is completely eliminated in the fully encapsulated device. (B) Chip micrograph and dimensions of the prototype ENIAC.

As shown in Fig. 8.2, the ENIAC is designed to be small enough ($3 \times 3 \times 0.25$ mm^3) to be placed among the folds and curves of the cortical surface (see Fig. 1.3), and to be implanted through small skull fissures. Hence it offers greater coverage of the cortical surface while being much less obtrusive than other minimally invasive ECoG approaches, permitting even insertion without surgery. As seen in the block diagram in Fig. 8.2, the chip contains an LC resonant tank, electrodes, recording channels, stimulator, power management units, and bidirectional communication circuits.

Its first prototype, fabricated in a 180-μm CMOS silicon-on-insulator (SOI) process, is shown on the upper right side of Fig. 8.2. With two turns and 100-μm thickness, the on-chip coil results in an inductance of 23.7 nH. The same single coil is shared for wireless power transfer and bidirectional RF communication. Sixteen electrodes, which can be individually configured as recording or stimulating channels, are integrated directly on the top metal layer of the chip. To enhance energy efficiency and remove the need for separate rectification and regulation stages, an integrated resonant regulating rectifier IR^3 [27] is implemented. In addition, an adiabatic stimulator generates constant-current stimulation pulses from the RF power input in an adiabatic manner, much more energy efficient than conventional stimulation from DC static power supplies.

8.3. Power and communication

As highlighted in Sect. 1.6.1, RF inductive powering is the most efficient means for power delivery at this implantation depth, and has been adopted in ENIAC. To model the inductive link through tissue, a detailed finite element method (FEM) model of the octagonal loop transmitter antenna and the 3×3-mm^2 ENIAC shown in Fig. 8.3(A) was constructed in ANSYS HFSS, using tissue spectral permittivity and absorption properties as shown in Fig. 1.7(B)–(C). Optimal power transfer between the transmitter coil and ENIAC is reached at a resonance frequency of 190 MHz as shown in Fig. 8.3(B). At this frequency, substantially more than the required 2 mW power can be delivered

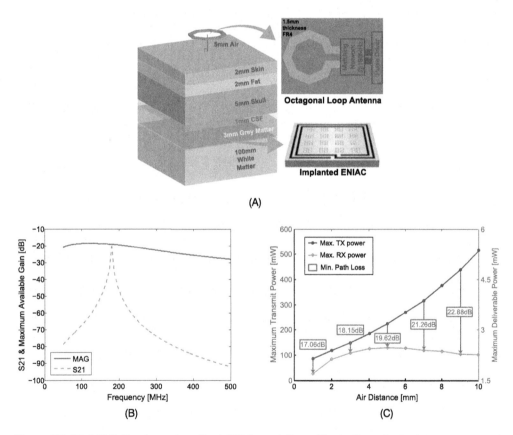

(A)

(B)

(C)

Figure 8.3 (A) 3-D finite element method (FEM) modeling of brain tissue layers between external transmitter and implanted ENIAC. (B) Simulated forward transmission coefficient S$_{21}$ and maximum available gain (MAG) from the transmitter to the implanted ENIAC. The optimal frequency for wireless power transfer is around 190 MHz. (C) Maximum transmit power limited by specific absorption rate (SAR) and maximum receivable power at the implanted ENIAC, as a function of distance of air gap between the transmitter and the scalp, optimum around 5 mm. Green arrows (mid gray in print version) denote minimum path losses at each air distance.

under the specific absorption rate (SAR) limit (2 W/kg in IEEE std. 1528). Fig. 8.3(C) shows the maximum transmit power at the SAR limit, and corresponding maximum deliverable power at the implant, for varying distance of the air gap between the loop transmitter and the scalp. The optimal distance for maximum power delivery, trading between reduced SAR-limited transmit power at lower distance and increased path losses at higher distance [28], was found to be around 5 mm.

ENIAC minimizes power losses in the received power from the RF coil owing to an integrated resonant regulating rectifier (IR^3) architecture that combines power management stages of rectification, regulation, and DC conversion, eliminating typical losses due to inefficiencies at each stage when implemented separately. As illustrated in Fig. 8.4(A), IR^3 generates a constant power supply 0.8 V independent of fluctuation in the LC tank voltages. IR^3 operates by adapting both width and frequency of pulsed rectifier switching based on a feedback signal derived from V_{DD} [27].

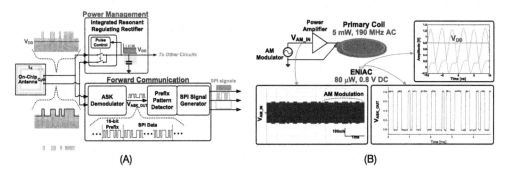

Figure 8.4 (A) System diagram of ENIAC power management and ASK forward communication, sharing the same single on-chip loop antenna. The integrated resonant regulating rectifier (IR^3) generates a stable 0.8 V DC output voltage directly from the 190 MHz RF coil voltage while the ASK demodulator decodes and amplifies the modulated signal. (B) Simplified test setup for wireless powering and communication along with measurement samples at the transmitter and the receiver.

Concurrently, the amplitude-shift-keying (ASK) demodulator tracks and amplifies the envelope of the LC tank voltages to decode transmitted configuration data as illustrated on the bottom of Fig. 8.4(A). The ASK communication is used to wirelessly configure the operation modes and parameters of the chip. To synchronize data reception, a 16-bit pre-determined identification code is used as prefix followed by serial peripheral interface (SPI) signals.

Fig. 8.4(B) shows test setup and sample data for the IR^3 power delivery and the ASK data transmission. For these tests, a primary coil built on a printed-circuit board was placed 1 cm above the ENIAC. The top right panel in Fig. 8.4(B) shows the measured coil voltages simultaneously rectified and regulated by the IR^3 [27,29] to produce the supply voltage $V_{DD} \approx 0.8$ V. The total transmitted power is about 5 mW, of which around 80 µW is received by the ENIAC. The bottom panel of Fig. 8.4(B) shows the AM modulated input on the primary side, and the demodulated ASK signal in the ENIAC.

8.4. Signal recording and stimulation

8.4.1 Recording

The recording module integrates 16 analog front-ends (AFEs), a 16:1 analog multiplexer (MUX), and an analog-to-digital converter (ADC) as shown in Fig. 8.5(A). Each of the 16 capacitively coupled electrodes is connected either to its local AFE channel, or to the global stimulator, multiplexed by a high-voltage tolerant switch matrix. The AFE amplifies the biopotential V_{IN1} from the capacitively coupled non-contact electrode with two amplification stages and a common-mode averaging circuit. The common-mode averaging circuit constructs a single reference signal V_{AVG} as the average of all V_{INi}

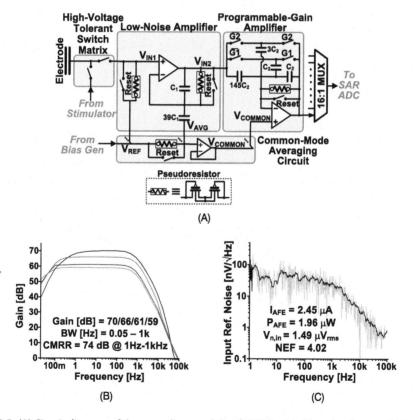

Figure 8.5 (A) Circuit diagram of the recording module of ENIAC with 16 analog frontend (AFE) channels, 16:1 analog multiplexer (MUX), and successive approximation register (SAR) analog-to-digital converter (ADC). (B) Measured frequency and noise characteristics of one AFE channel.

electrode voltages through capacitive division. Similar to differential recording across a pair of adjacent electrodes (Sect. 1.3.1), the internal common-mode reference V_{AVG} allows single-ended recording over all 16 electrodes without the need for a distal external ground connection. A pMOS-based pseudoresistor (in the inset of Fig. 8.5(A)) is used to set the DC operating point at V_{REF} for the capacitive division to allow for very high (TΩ-range) resistance in very small silicon area [30,31].

The first low-noise amplifier stage has a non-inverting configuration with a feedback capacitor C_1 and a common-mode coupling capacitor of $39 \cdot C_1$, which connects to the common-mode averaging node V_{AVG}, resulting in a differential voltage gain of 40 (V/V). V_{AVG} is buffered and used for common-mode rejection in the second AFE stage. The second AFE stage provides variable gain by manipulating the connections of two capacitors, connected either as input or as feedback capacitors [32,33]. Output signals of the AFEs are multiplexed and buffered to the SAR ADC, which has time-interleaving sample-and-hold input DACs to ensure longer sampling time, leading to power saving in buffering the input DAC of the ADC.

Measurement results for the AFE, characterizing its frequency response and noise performance, are shown in Fig. 8.5(B)–(C). Variable 50–70 dB gain is supported, and the input-referred noise is 1.5 μV_{rms} at 2.45 μA supply current for a noise efficiency factor (NEF) of 4.

8.4.2 Stimulation

As illustrated in Fig. 8.6(A), ENIAC on-chip electrodes are implemented on top metal, as used for bond pads and on-chip inductors. The exposed electrodes allow for direct coating with a thin film of high-k materials such as TiO_2/ZrO_2 to achieve high capacitance for high charge delivery capacity. With 30-nm coating and 250×250 μm^2 area, the coupling capacitance C_{EL} is about 1.5 nF, one order of magnitude smaller than that of a platinum electrode of same area. Total deliverable charge per phase Q_{ph} can be expressed as

$$Q_{ph} = I_{STM} \cdot T_{ph} = C_{EL} \cdot V_{DD\_STM} \tag{8.1}$$

where I_{STM} is the stimulation current, T_{ph} the time duration of the phase, and V_{DD_STM} the total voltage dynamic excursion. Relatively low capacitance C_{EL} can thus be compensated by an increased total voltage excursion V_{DD_STM} to deliver the required charge per stimulation phase. In order to achieve $Q_{ph} = 10$ nC per stimulation phase, needed for effective neural stimulation under typical electrophysiological conditions, a dynamic voltage rail with a total excursion of more than 8 times the static supply voltage V_{DD} (= 0.8 V) is required.

Conventionally, this can be implemented by generating the required high power supply voltages and supplying constant currents from fixed power rails. However, draw-

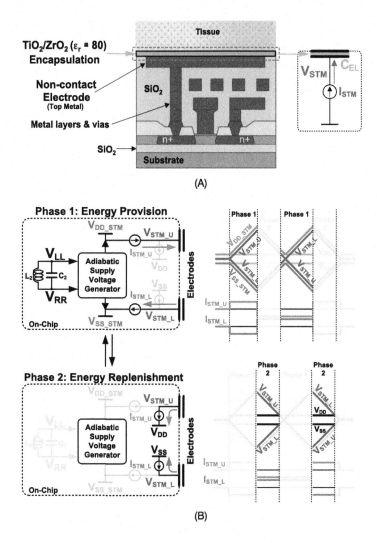

Figure 8.6 (A) Simplified stack-up of ENIAC showing an electrode coated with high-*k* materials for capacitive interface. (B) Principle of adiabatic stimulation with ENIAC. During the first phase, adiabatic voltage rails are generated directly from the LC tank for energy efficient stimulation. During the second phase, the energy stored across the capacitive electrodes is replenished for further energy savings.

ing currents in this manner incurs large energy penalties due to the large voltage drop across the current source.

Instead, a much better way to perform stimulation is to slowly ramp up the supply rails in an adiabatic fashion to minimize the voltage drop across the current source. Generation of the adiabatic voltage rails can be implemented in various ways. External capacitors [37] or an external inductor [38] can be employed. Alternatively, pulse width control in rectifier can be used [39]. However, all of these methods have output ranges within the LC tank swing voltages or V_{DD}. Recently, on-chip charge pumps were employed to generate a wide voltage excursion for adiabatic stimulation [40]. Because this approach utilizes the DC power supply as the input of charge pumps, series of power efficiency loss cannot be avoided in implantation settings. In addition, this method could generate discrete levels of power supplies only, so the energy losses due to the voltage drops across the current source were considerable.

In contrast, ENIAC implements an adiabatic stimulator that generates, at minimum energy losses, ramping voltage power rails with greater than 8 times the voltage excursion of the LC tank, and with no need for any external components. Consistent with the observations in Sect. 1.3.2, differential adiabatic stimulation across a selected pair of electrodes is implemented, since the miniaturized and enclosed ENIAC system permits no access to a distal ground electrode. As illustrated in Fig. 8.6(B), the ENIAC stimulator operates in two phases. During the first phase, constant complementary currents are provided through the differential capacitive electrodes. The ramping voltage adiabatic power rails V_{DD_STM} and V_{SS_STM} providing the complementary currents are generated directly from the LC tank utilizing a foldable stack of rectifiers. During the second phase energy is replenished, by returning the charge stored on the electrode capacitors to the system V_{DD} and V_{SS} for reuse by other ENIAC modules. For triphasic rather than biphasic stimulation, as shown, the two phases are repeated but now with opposite polarity. This is accomplished by swapping the electrode connections through the switch matrix prior to executing the same two-phase sequence. Finally, the electrodes are shorted to even out any residual charge on the electrode capacitors.

Fig. 8.7 shows measured voltage and current waveform for the triphasic stimulation with a platinum model electrode, consistent with the model in Fig. 8.6(B), and showing 145 μA of current delivered per electrode channel. This is 4 times larger than other

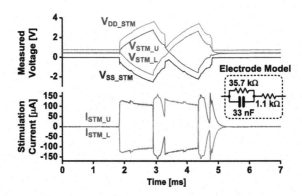

Figure 8.7 Measured stimulation voltage and current waveforms with platinum model electrode.

Table 8.1 Comparison of state-of-the-art wireless integrated ECoG recording and stimulation systems.

Reference		[34]	[35]	[20,25] WIMAGINE	[36,22]	This work ENIAC
Number of channels		8[a]	64	32	64	16
Technology		0.18 μm	0.18 μm	0.35 μm	65 nm	0.18 μm SOI
Power supply [V]		1.8	1.8	3.3	0.5	0.8
Total power consumption [mW]		2.8	5.4[b]	72.1	0.22	<0.1
IC area [mm²]		13.47	26.83[b]	86[b]	5.76	9
Total system volume [mm³]		N/A	77,200[c]	>4000	N/A	2.25
External components		LC, electrodes; Antenna; Capacitor	LC, electrodes; Antenna; Off-chip ICs for powering and communication	LC, electrodes; Antenna	L, electrodes; Antenna	None
AFE	Input Ref. noise [μV]	5.2	5.4	1	1.3	2.5
	NEF	1.8	N/A	4.5	4.8	4.0
ADC	Type	SAR	SAR	SAR	VCO	SAR
	Resolution [bits]	10	12	12	15	10
Stimulation	Max. output current [μA]	30	No	No	No	145
Wireless powering	Frequency [MHz]	13.56	0.266	13.56	300	190
	Antenna	Off-chip	Off-chip	Off-chip	Off-chip	On-chip
Communication	Forward	OOK 401–406 MHz	Zigbee N/A	No	No	ASK 190 MHz
	Backward	OOK 401–406 MHz	IR–UWB 7.3–8.5 GHz	FSK 402–405 MHz	LSK 300 MHz	LSK 190 MHz

[a] 8 recording channels and 2 stimulation channels.
[b] Only for the data acquisition unit.
[c] Only for the communication unit.

integrated ECoG systems even though no external components are used and system volume is substantially smaller (Table 8.1).

8.5. Conclusions

In this chapter we make the case for a new type of device that promises to expand the applications of implantable brain monitoring – free-floating modular μECoG. We demonstrate a new approach to miniaturization of modular μECoG with our fully integrated Encapsulated Neural Interfacing Acquisition Chip (ENIAC). This system-on-a-chip is capable of recording, stimulating, wireless power conditioning and bidirectionally communicating without the need for any external components. Its major specifications, performances and functionalities are summarized in comparison with other state-of-the-art ECoG interface systems in Table 8.1. Having a fully integrated neural interface system, including electrodes and antenna, is a new milestone for miniaturization that sets the stage for exciting clinical and research developments.

References

[1] F.T. Sun, M.J. Morrell, R.E. Wharen, Responsive cortical stimulation for the treatment of epilepsy, Neurotherapeutics 5 (1) (2008) 68–74.

[2] NeuroPace RNS system, [Online]. Available: http://www.neuropace.com/.

[3] D. Prutchi, Neurovista publishes study results for their implantable seizure-warning device, [Online]. Available: http://www.implantable-device.com.

[4] K.A. Davis, B.K. Sturges, C.H. Vite, V. Ruedebusch, G. Worrell, A.B. Gardner, K. Leyde, W.D. Sheffield, B. Litt, A novel implanted device to wirelessly record and analyze continuous intracranial canine EEG, Epilepsy Research 96 (1–2) (2011) 116–122.

[5] M.J. Cook, T.J. O'Brien, S.F. Berkovic, M. Murphy, A. Morokoff, G. Fabinyi, W. D'Souza, R. Yerra, J. Archer, L. Litewka, S. Hosking, P. Lightfoot, V. Ruedebusch, W.D. Sheffield, D. Snyder, K. Leyde, D. Himes, Prediction of seizure likelihood with a long-term, implanted seizure advisory system in patients with drug-resistant epilepsy: a first-in-man study, Lancet Neurology 12 (6) (2013) 563–571.

[6] A.G. Rouse, S.R. Stanslaski, P. Cong, R.M. Jensen, P. Afshar, D. Ullestad, R. Gupta, G.F. Molnar, D.W. Moran, T.J. Denison, A chronic generalized bi-directional brain–machine interface, Journal of Neural Engineering 8 (3) (2011) 036018.

[7] P. Afshar, A. Khambhati, S. Stanslaski, D. Carlson, R. Jensen, D. Linde, S. Dani, M. Lazarewicz, P. Cong, J. Giftakis, P. Stypulkowski, T. Denison, A translational platform for prototyping closed-loop neuromodulation systems, Frontiers in Neural Circuits 6 (2013).

[8] M. Hirata, K. Matsushita, T. Suzuki, T. Yoshida, F. Sato, S. Morris, T. Yanagisawa, T. Goto, M. Kawato, T. Yoshimine, A fully-implantable wireless system for human brain–machine interfaces using brain surface electrodes: W-HERBS, IEICE Transactions on Communications E94b (2011) 2448–2453.

[9] M. Hirata, T. Yoshimine, Electrocorticographic brain–machine interfaces for motor and communication control, in: K. Kansaku, L.G. Cohen, N. Birbaumer (Eds.), Clinical Systems Neuroscience, Springer Japan, 2015, pp. 83–100, ch. 5.

[10] F.T. Sun, M.J. Morrell, The RNS system: responsive cortical stimulation for the treatment of refractory partial epilepsy, Expert Review of Medical Devices 11 (6) (2014) 563–572.

[11] C.N. Heck, D. King-Stephens, A.D. Massey, D.R. Nair, B.C. Jobst, G.L. Barkley, V. Salanova, A.J. Cole, M.C. Smith, R.P. Gwinn, C. Skidmore, P.C. Van Ness, G.K. Bergey, Y.D. Park, I. Miller, E. Geller, P.A. Rutecki, R. Zimmerman, D.C. Spencer, A. Goldman, J.C. Edwards, J.W. Leiphart, R.E. Wharen, J. Fessler, N.B. Fountain, G.A. Worrell, R.E. Gross, S. Eisenschenk, R.B. Duckrow, L.J. Hirsch, C. Bazil, C.A. O'Donovan, F.T. Sun, T.A. Courtney, C.G. Seale, M.J. Morrell, Two-year seizure reduction in adults with medically intractable partial onset epilepsy treated with responsive neurostimulation: final results of the RNS system pivotal trial, Epilepsia 55 (3) (2014) 432–441.

[12] M. Schuettler, F. Kohler, J.S. Ordonez, T. Stieglitz, Hermetic electronic packaging of an implantable brain–machine-interface with transcutaneous optical data communication, in: Proceedings of the Annual International Conference of the IEEE Engineering in Medicine and Biology Society, 2012, pp. 3886–3889.

[13] F. Kohler, M.A. Ulloa, J.S. Ordonez, T. Stieglitz, M. Schuettler, Reliability investigations and improvements of interconnection technologies for the wireless brain–machine interface – 'BrainCon', in: Proceedings of the International IEEE/EMBS Conference on Neural Engineering, 2013, pp. 1013–1016.

[14] J.D. Fischer, The Braincon Platform Software – A Closed-Loop Brain–Computer Interface Software for Research and Medical Applications, 2014.

[15] High-density microelectrocorticography array, [Online]. Available: http://corteraneuro.com/products/neuroscience.

[16] R. Mohammadi, M.A. Sharif, A. Kia, M. Hoveidar-Sefid, A.M. Sodagar, E. Nadimi, A compact ECoG system with bidirectional capacitive data telemetry, in: Proceedings of the IEEE Biomedical Circuits and Systems Conference, 2014, pp. 600–603.

[17] Y.-C. Chen, H.-L. Hsu, Y.-T. Lee, H.-C. Su, S.-J. Yen, C.-H. Chen, W.-L. Hsu, T.-R. Yew, S.-R. Yeh, D.-J. Yao, Y.-C. Chang, H. Chen, An active, flexible carbon nanotube microelectrode array for recording electrocorticograms, Journal of Neural Engineering 8 (3) (2011) 034001.

[18] G.A. DeMichele, S.F. Cogan, P.R. Troyk, H. Chen, Z. Hu, Multichannel wireless ECoG array ASIC devices, in: Proceedings of the Annual International Conference of the IEEE Engineering in Medicine and Biology Society, 2014, pp. 3969–3972.

[19] A. Ghomashchi, Z. Zheng, N. Majaj, M. Trumpis, L. Kiorpes, J. Viventi, A low-cost, open-source, wireless electrophysiology system, in: Proceedings of the Annual International Conference of the IEEE Engineering in Medicine and Biology Society, 2014, pp. 3138–3141.

[20] S. Robinet, P. Audebert, G. Regis, B. Zongo, J.F. Beche, C. Condemine, S. Filipe, G. Charvet, A low-power 0.7 μV_{rms} 32-channel mixed-signal circuit for ECoG recordings, IEEE Journal on Emerging and Selected Topics in Circuits and Systems 1 (4) (2011) 451–460.

[21] M. Schuettler, F. Kohler, J.S. Ordonez, T. Stieglitz, Hermetic electronic packaging of an implantable brain–machine-interface with transcutaneous optical data communication, in: Proceedings of the Annual International Conference of the IEEE Engineering in Medicine and Biology Society, 2012, pp. 3886–3889.

[22] R. Muller, H.-P. Le, W. Li, P. Ledochowitsch, S. Gambini, T. Bjorninen, A. Koralek, J.M. Carmena, M.M. Maharbiz, E. Alon, J.M. Rabaey, A minimally invasive 64-channel wireless μECoG implant, IEEE Journal of Solid-State Circuits 50 (1) (2015) 344–359.

[23] J. Viventi, D.H. Kim, L. Vigeland, E.S. Frechette, J.A. Blanco, Y.S. Kim, A.E. Avrin, V.R. Tiruvadi, S.W. Hwang, A.C. Vanleer, D.F. Wulsin, K. Davis, C.E. Gelber, L. Palmer, J. Van der Spiegel, J. Wu, J.L. Xiao, Y.G. Huang, D. Contreras, J.A. Rogers, B. Litt, Flexible, foldable, actively multiplexed, high-density electrode array for mapping brain activity *in vivo*, Nature Neuroscience 14 (12) (2011) 1599–1605.

[24] T.J. Richner, S. Thongpang, S.K. Brodnick, A.A. Schendel, R.W. Falk, L.A. Krugner-Higby, R. Pashaie, J.C. Williams, Optogenetic micro-electrocorticography for modulating and localizing cerebral cortex activity, Journal of Neural Engineering 11 (1) (2014) 016010.

[25] C.S. Mestais, G. Charvet, F. Sauter-Starace, M. Foerster, D. Ratel, A.L. Benabid, WIMAGINE: wireless 64-channel ECoG recording implant for long term clinical applications, IEEE Transactions on Neural Systems and Rehabilitation Engineering 23 (1) (2015) 10–21.

[26] S. Ha, A. Akinin, J. Park, C. Kim, H. Wang, C. Maier, G. Cauwenberghs, P.P. Mercier, A 16-channel wireless neural interfacing SoC with RF-powered energy-replenishing adiabatic stimulation, in: Symposium on VLSI Circuits Digest of Technical Papers, 2015, pp. C106–C107.

[27] C. Kim, S. Ha, J. Park, A. Akinin, P.P. Mercier, G. Cauwenberghs, A 144 MHz integrated resonant regulating rectifier with hybrid pulse modulation, in: Symposium on VLSI Circuits Digest of Technical Papers, 2015, pp. C284–C285.

[28] J. Jian, M. Stanaćević, Optimal position of the transmitter coil for wireless power transfer to the implantable device, in: Proceedings of the Annual International Conference of the IEEE Engineering in Medicine and Biology Society, 2014, pp. 6549–6552.

[29] C. Kim, J. Park, A. Akinin, S. Ha, R. Kubendran, H. Wang, P.P. Mercier, G. Cauwenberghs, A fully integrated 144 MHz wireless-power-receiver-on-chip with an adaptive buck-boost regulating rectifier and low-loss H-tree signal distribution, in: Symposium on VLSI Circuits Digest of Technical Papers, 2016.

[30] T. Delbruck, C.A. Mead, Adaptive photoreceptor with wide dynamic range, in: Proceedings of the IEEE International Symposium on Circuits and Systems, vol. 4, 1994, pp. 339–342.

[31] R.R. Harrison, C. Charles, A low-power low-noise CMOS amplifier for neural recording applications, IEEE Journal of Solid-State Circuits 38 (6) (2003) 958–965.

[32] X. Zou, X. Xu, L. Yao, Y. Lian, A 1-V 450-nW fully integrated programmable biomedical sensor interface chip, IEEE Journal of Solid-State Circuits 44 (4) (2009) 1067–1077.

[33] D. Han, Y. Zheng, R. Rajkumar, G.S. Dawe, M. Je, A 0.45 V 100-channel neural-recording IC with sub-μW/channel consumption in 0.18 μm CMOS, IEEE Transactions on Biomedical Circuits and Systems 7 (6) (2013) 735–746.

[34] W.-M. Chen, H. Chiueh, T.-J. Chen, C.-L. Ho, C. Jeng, M.-D. Ker, C.-Y. Lin, Y.-C. Huang, C.-W. Chou, T.-Y. Fan, M.-S. Cheng, Y.-L. Hsin, S.-F. Liang, Y.-L. Wang, F.-Z. Shaw, Y.-H. Huang, C.-H. Yang, C.-Y. Wu, A fully integrated 8-channel closed-loop neural-prosthetic CMOS SoC for real-time epileptic seizure control, IEEE Journal of Solid-State Circuits 49 (1) (2014) 232–247.

[35] H. Ando, K. Takizawa, T. Yoshida, K. Matsushita, M. Hirata, T. Suzuki, Multichannel neural recording with a 128 Mbps UWB wireless transmitter for implantable brain–machine interfaces, in: Proceedings of the Annual International Conference of the IEEE Engineering in Medicine and Biology Society, 2015, pp. 4097–4100.

[36] T. Bjorninen, R. Muller, P. Ledochowitsch, L. Sydanheimo, L. Ukkonen, M.M. Maharbiz, J.M. Rabaey, Design of wireless links to implanted brain-machine interface microelectronic systems, IEEE Antennas and Wireless Propagation Letters 11 (2012) 1663–1666.

[37] S.K. Kelly, J.L. Wyatt, A power-efficient neural tissue stimulator with energy recovery, IEEE Transactions on Biomedical Circuits and Systems 5 (1) (2011) 20–29.

[38] S.K. Arfin, R. Sarpeshkar, An energy-efficient, adiabatic electrode stimulator with inductive energy recycling and feedback current regulation, IEEE Transactions on Biomedical Circuits and Systems 6 (1) (2012) 1–14.

[39] U. Çilingiroğlu, S. İpek, A zero-voltage switching technique for minimizing the current-source power of implanted stimulators, IEEE Transactions on Biomedical Circuits and Systems 7 (4) (2013) 469–479.

[40] W. Biederman, D.J. Yeager, N. Narevsky, J. Leverett, R. Neely, J.M. Carmena, E. Alon, J.M. Rabaey, A 4.78 mm^2 fully-integrated neuromodulation SoC combining 64 acquisition channels with digital compression and simultaneous dual stimulation, IEEE Journal of Solid-State Circuits 50 (4) (2015) 1038–1047.

Index

Printed in the United States
By Bookmasters